FOREWORD

10.22334. A grain of sand.

Variothoughts is sand, not gold. According to Syntalism, our world is built of sand, not gold. The most valuable thing is sand, sand is bread, and gold is salt.

10.22742.

Variothoughts books should be read slowly, chewing every thought carefully. Truth is that which has extension properties, and falsehood is pride, that is, an avid rush.

10.22339.

There are no questions that cannot be answered in the Variothoughts. The Variothoughts is an endless source of inspiration for hearts searching for truth.

10.22266. Nutcracker.

Breaking stereotypes and patterns. Variothoughts is a brain-crushing book, the meaning of which is to achieve the integrity of the mind. First, all beliefs should be destroyed in the dust, and then it will all stick together and enlightenment will come.

10.22554. The living and the dead book.

In the original, Variothoughts is the ideal of the perfection of truth, but the ideal is dead and, therefore, there is no joy in it. To bring back the joy of life to Variothoughts, I decided to salt the bread. Salt-free bread is too sweet. In Variothoughts translations, I threw a couple of spoons of chaos. My act of monstrous vandalism led to the loss of 20% of the meaning, and made the texts very strange and obscure. You will call me a scoundrel and a vandal, but I don't think so ... On the contrary, I believe it made the texts charming, created artificial barriers ... Now, to understand the texts of Variothoughts and find the truth, you have to smash your head and think. Thinking is joyful.

10.15530.

The basis of speech is truth. Cognition of truth should begin with clarifying the meaning of words.

10.16833.

The main feature of Variothoughts is its unprecedented honesty. No censorship of thoughts, absolute freedom of ideas and words.

5.412.

Variothoughts is a book for those who save their time. Ready-made Lego cubes used to put together any ideas and goals. The DNA and RNA of thought.

3.2212. Attainment of truth.

A comprehensive attainment of reality occurs by knowledge's thinning of its tiniest degree of detail.

3.2213.

Unity is hidden in differences. You unite by disuniting. Holding one onto another, the infinitely small becomes the infinitely big.

10.6168.

The Variothoughts is a basic library of DNA of thought, loading it into the brain can solve any problem. Any dreams, any goals will be available to you, thanks to the philosophy of Syntalism.

3.1971.

Variothoughts implements divergent thought algorithms in

order to come again to unity through a multitude. Many grows out of one and one grows again out of many.

3.2211.

Soundness is the ability to dynamically examine things from different perspectives.

3.2216.

Having reached its limit, knowledge transforms into will. What is the limit of knowledge? – Faith.

10.21243. Love of truth.

The meaning of human life is to overcome infinite loneliness and find infinite love.

THE EVOLUTION OF MAN AND THE WORLD. SYNTALIST THEORY OF EVOLUTION.

10.6878. The virus of evolution.

It is not a person who is evolving, but viruses and problems that want to devour him. Fighting viruses, a person learns about them, incorporates their DNA into their DNA, and defeats viruses with their own weapons. Altered human DNA changes human nature. Viruses fall back and come up with a new weapon, attack again. Thus, the more problems a person solves, the more he mutates and evolves.

10.6879.

The reason for evolution is viruses, which, in fact, are elements of inanimate nature. Thus, the factor of inanimate (pseudo-living nature) is the main engine of life. In fact, inanimate nature is evolving in order to destroy the living.

10.6978. Tell me who your enemy is and I will tell you who you are.

To grow and become stronger, you need enemies. Knowing your enemies, you take their weapons. Your enemies are the source of your power and energy. A person's DNA is 30% made up of the

DNA of viruses that wanted to harm them.

10.6985. The evolutionary factor.

When we fall in love, we take something unique, something that we don't have. Everyone has the same thing in common. Unique is an element of evolution. Love allows us to evolve and acquire new, hitherto unknown properties.

10.7071. An exotic monkey.

It is a lie that man is descended from an APE. Man is an APE, just one of hundreds of breeds of apes.

10.7093.

Why don't apes evolve into humans? For the same reason that a monkey doesn't turn into a chimpanzee or a gorilla into a baboon. Man is one of the breeds of apes. Previously, there were different breeds of apes of the human species, but then they became extinct or they were assimilated by the main species of homosapiens.

10.7094. Hobbits.

On Paradise island, the height of a person has decreased to one meter, and the brain has shrunk to 600 grams. When there are no enemies and problems, brains and strength are not needed.

10.7128.

Conflict and war are the engines of evolution. Predators, suffering from hunger, try to eat their victims, and the victims, knowing the predators, learn to defeat them.

10.7163. The dam is overflowing.

In many ways, the cause of human evolution was pathogenic

viruses and bacteria. Medicine, defeating diseases, not only stops evolution, it also weakens the immune system. In parallel with the weakening of the person, the diseases become stronger and stronger.

10.7642.

Man does not evolve by himself. His desires evolve, his desires create problems. Overcoming problems, man evolves. The more serious your enemies and problems, the stronger you become.

10.8490.

The uniqueness of the human mind is that it can purposefully create objects that never existed.

10.9201.

Let's be honest, people settled around the world and evolved not from a good life, but rather from despair. The question was: either progress and evolution, or death. The good life only degrades people.

10.9900.

The most successful predator on planet Earth is viruses. Every day, viruses kill half of the world's bacterial population, but they do not give up and are born again. The war of viruses and bacteria generates the wave function of life. Man in this struggle is like mushrooms. Mushrooms in this system are the third force that stabilizes and maintains the balance. We can't allow either side to win, if the war ends, it will destroy life.

10.12108.

Redundancy generates adaptability and variability.

10.12289.

People are said to have descended from apes.
– I'll tell you more, the first animals were worms. We are all descended from tapeworms.

10.12417.

Brains were created to lie and deceive predators.

10.13439.

The monkey turned into a man because it was kicked out of the house. There was not enough wood for everyone. Those monkeys that remained in the forest remained macaques.

10.16822.

As the Bible teaches us, man was created by God... According to the development of this thought, God took a brainless monkey and gave it brains. "Let there be light of reason!"God said ... and so it all gradually spun.

10.18712.

It is interesting that animals adapt to the environment of existence, and intelligent beings adapt the environment of existence to themselves.

10.19037.

The essence of natural selection is to separate the grain from the chaff. The tares will perish in the trap of vices, and the grains will leave them unnoticed.

10.19650.

The point is to learn to eat what others can't.

10.20902.

The ecological niche is very dependent on predators and parasites. To find your place in life, you need to find the predator that will devour you. Strong predators protect their food from other predators.

10.21021.

In the evolutionary struggle, the weak one wins, because the poor in spirit are blessed and ready to love their enemy. The fact that predators protect their victims, predators benefit from reproduction and the safety of their food. The sheep who have placed themselves in the power of a prudent shepherd fall under his iron protection.

10.21146.

We should not be afraid of those who eat us. The one who wants to eat us is our enemy, and therefore our friend.

10.21169. Coral Paradise.

Do not rush to devour your victims, because a predator that is restrained in killing its victims is useful for them, because it protects them from other predators. Thus, the victim itself will seek to enslave the slow and caring predator.

10.21286.

Weakness is strength, because in the evolutionary struggle the weakest wins.

10.21374.

You are tormented by thoughts about the uselessness of being, because you do not know that the horses go in front and the cart behind.

10.21420. Nature is a wonderful cook.

Wormy Apple is better not a worm-eaten Apple. Wormy Apple is a delicacy, salad with meat. Animals are very fond of this delicacy.

10.21421.

Parasites are useful because they are very hungry for food and life, so it is profitable for them to protect their victims from other predators, as well as to promote their reproduction.

10.22142. Thirteen billion years of evolution.

God took his time. First he created the earth for six days, then he rested for a day... do you Think he's satisfied with that?

7.82.

The work created from the APE man, and the horse is not created. That's because the horse thinks too much.

9.52. Evolution of human into a radio-controlled ant.

The technology of an electronic assistant based on artificial intelligence, which will help human to remember information and suggest something, will most likely lead to total degradation of the human brain, which will quickly destroy humanity. Now, when it`s not necessary to think, the brains will be reduced even more and human will turn into an insect. The assistant will manage human through a radio channel, and human, as it befits an ant, will crawl and collect energy, dragging it in the anthill. The more energy the ant gets, the happier it will be.

9.80.

Viruses and diseases kill the weak and the identical. The essence of this phenomenon is the evolution and natural selection of perfect and unique individuals.

1439. Evolution of cybernetic systems...

Once the world was ruled by dinosaurs,
then predators,
people came for them,
but....
Everything, as you know, passes:
kings, emperors,
the oligarchs, presidents
Ty.
The time has come for corporations,
in essence its,
cold cybernetic systems,
soulless cyborgs.

cyborgs rule the world.
They were created to grow.
Inside them are little men-dwarfs
twist the gear wheel
huge machines.
Huge soul of the cyborg
knows only one goal:
...endless, insane growth,
until everything is consumed.
Less perfect forms of life-people...
- become superfluous...

1705. Sex, relationship and chemistry.

It is considered to be rational to take a shower and wash oneself before sex.
Generally, it's really rational but there are some details...
Everyone has one's own personal smell. The chemistry of the body. And this smell depends on many things, like hormones, glands etc. And this smell is what differs a particular person among million others.

A touch of civilization for the last ten thousand years hasn't drastically changed human nature. As evolution is a matter of million years. And ten thousand years don't bring any changes. And animals, as is known, attach great importance to smells. They smell each other to realize whether it's their type of creature.

Thus, when a male and a female meet for the first time and decide whether they are a nice match, their decision also depends on (or maybe largely depends on) the smell... What's more, as soon as they too become a couple they start to produce some identical and exactly mutually suitable pheromones that fit like a lock and the key to it. That very signal "my type- another type". All these pheromones, smells etc. form a chemical bond between a male and a female and bring them pleasure while making them closer and loyal to each other as other partners no longer represent interest for them.

Thus, when one washes all one's smells and pheromones off and starts to smell of soap or fragrance, it may spoil the chemical base of a relationship and deprive people of big "animalistic pleasure".

Just think of Napoleon and Josephine... Five days before he visited her, Josephine didn't wash so that when Napoleon visited her he simply lost his mind with wild desires... Perhaps five days is a bit too many, but however, it's an obvious historical fact...

1829.

You want to protect a man from himself? Why? Do you feel like a God and want to prevent evolution from doing its job?

2272. A spiral of evolution.

The rise and fall of humankind are two different phenomena.

1% of people evolve and become «Gods».
99% evolve and "turn into bees, ants, aphides, wasps, worms and flies»...

At first it will cause a rise but in time, Gods who don't face difficulties and competition, will fall. The rise will end...and a long night will come. Later, in a million years, there will be a new rise and a new spiral turn.

2614.

Thinking is good. The presence of traces of thought in the head distinguishes a person from a monkey.

2615. Key difference.

Monkey thinks how to get banana from the palm.
While a man thinks whom he can make to get banana from the palm.

3.367. Parents' pride.

The meaning of the conflict between fathers and children is that you can not grow above the one from whom you take an example. However, life requires us to evolve and grow, so children, taking everything they can from their parents, reject and choose a higher goal for growth. Parents not in forces to transfer such disobedience, full of anger, pride, resentment and jealousy fall into infamy.

3699.

The Following development of evolution can be presumed: silicon, iron, nickel an magnet field(water analogue)-based form of life created hydrocarbon and electricity based form of life, form of life, which, in its turn, created silicon and electricity based form of life. On the other hand, probably a way to a silicon-electric form of life is a kind of regress.

4198. Inevitability.

In essence, Drake equation that estimates the number of extraterrestrial civilizations, is a right thing. However, the problem is that any civilization that achieves a certain progress, soon becomes doomed to death.

The transition into digital state of existence. The evolution of a human being into a stone is the latter end of any intelligent life. They simply pass from the material world into some higher digital level, or decode the digital code of the Universe and connect to it in order to dissolve in it. That's why mankind is unable to find any evident signs of other intelligent civilizations.

4239.

A woman is the main engine for evolution...
It's their task to choose best men for propagation, namely the ones who possess novelty and perfection. She who chooses the right man as the father for her children, will be loved by destiny.

4362. Aging program.

The issue of aging and death of the body is a question of a special software package programmed for a certain mode of operation and is directly related to active sex life and reproduction, as well as the evolution and adaptability of mankind.

In other words, a person could live many orders of magnitude longer and his biological systems are capable of it, the question is-why? Of course, it would be possible to reprogram this program, but this lengthening of life is contrary to the idea of evolution. This solution will significantly slow down and even stop evolution, and, most likely, even destroy man as a species, causing a General system failure.

The fact is that life is a very complex software complex, in which millions of programs work, connected by a single goal. Changing such an important system variable as the rate of aging and the lifetime of an individual can lead to the collapse of the system. Stop normally work all instincts and laws. Society will cease to be renewed, the institution of the family and the whole socialization and economy of human society will be destroyed. People simply will not be able multiply, children and parents start kill each other, trying to preserve or obtain control of energy. Humanity will lose its ability to adapt and adapt.

In other words, to climb into the settings of system programs and human functions is fraught with a lot of problems.

4.441.

The best of sinners are still sinners. The most beautiful monkey is still a monkey. Beauty doesn't change the nature of things.

5.926. Corals are dying today.

The development and rise of human civilization can be attributed to the appearance of about 10 thousand years ago on our planet coral reefs, which concentrated in shallow coastal waters of fish and shellfish, providing the human mind with additional squirrel, previously inaccessible to him. An additional energy resource has allowed humanity to make a civilizational and technological breakthrough. Even the first money was shells.

6.660.

Philosophy is not taught in public school, because a grenade in the hands of a monkey is a terrible weapon.

7.203. The insects' is even smaller.

Not all humans evolved from apes – reptiloids evolved from reptiles. The reptilians' key feature is the extremely small size of their brain as compared to that of mammals.

7.216.

From nature's viewpoint, obesity is dangerous to monkeys: a branch may break off with their weight, and the monkey will fall to death.

7.219.

Some women are not descended from monkeys, but from gods and are called goddesses. For normal people it is better to avoid such women, because interspecies sex is a perversion.

7.220. A distant relative of God.

Humble people agree that their distant ancestors were apes. Vain and full of sinful pride people believe that they are descended from the Gods.

7.675. Eternal fear and hunger.

Homo sapiens monkeys managed to develop their intelligence and become humans because their ecological niche had been destroyed and, to survive, they had to start thinking. Other monkey species have their own ecological niches that are safe and full of food, so they feel no need to develop their intelligence.

7.710. On the edge of life.

We need integrity, we need to think and do at the same time. Monkeys always lived on the very edge of trees, jump and think need was almost simultaneously, and otherwise fall and death. It's when you think and do at the same time.

7.979. A universal curve.

Evolution engenders revolution. It is depicted graphically by an exponential curve, first really, really slowly, and then, after it achieves critical mass, blast and boost

7.987.

A good memory cans a thought and, finally, leads to its deterioration and death. The opportunity to forget is an extremely useful opportunity, which opens the way for evolution and all things new.

8.104. Society.

Little monkeys like to imitate everything that big monkeys do.

8.174.

Being successful when you're perfect and good is easy, but what about trying to survive and succeed when you're full of faults.

8.595.

It's possible to change anything but it's a matter of time and evolution.

9.157. Life arose in the mud.

Oxygen life arose in mud boilers (volcanoes) and only then got into the ocean. I would assume that life infected the ocean and arose on land. Nitrogen bases consumed ultraviolet light, released heat, warming up the atmosphere. This caused warming and melted the ice, raising the level of the oceans. Having sunk mud pots, the ocean got contracted with life. The first billion years there was life based on iron photosynthesis in the ocean and its extinction was associated with the exhaustion of resources.

9.257.

Immortality of individuals will stop evolution and reduce adaptability to the changing conditions of life. As a result, the population will be destroyed by negative circumstances, which it will not be able to adapt to. For example, evolving, viruses will adapt to the immune system of the immortals and wipe them off the face of the Earth.

9.305. Stop of the motion.

The problem of gods is their immortality. That was the very immortality that killed all the gods. Immortality stops evolution and turns the adaptation mechanisms off, while hostile circumstances, viruses, fungi and bacteria, keeping evolving, hack their immobilized victim and kill it.

9.407.

The Moon is the satellite that highly positively affects life on the planet. Adequate in size, the Moon can, by gravitational fluctuations, warm bowels of the planet around which it revolves, provoking volcanic activity, the basis of the atmosphere and life.

9.464. Evolution of food.

At first they were eating plankton, then small fishes, then insects, then climbed a tree and became fruit-eating monkeys, then got off the tree and started eating any and everything. But it's meat that turned them into human.

9.480. Coincidence.

The basis of life on the Earth include: distance to the Sun, size and content of the lithosphere, volcanic activity, meteors and water, the Moon warming bowels and generating volcanism.. And a range of factors linked with balance of iron, magnesium, sulphur and oxygen in the atmosphere, bacteria and plants waste products.

9.516. Victory of the intellect over biology.

The evolution of the human brain and human philosophy can be compared with the evolution of computers and mathematical algorithms that they compute. Over the past 15 years, computers have become 1000 times faster and algorithms – 29000 times faster. In total, this gave an increase in the operational speed by 29 million times (a little later this number reached 450 billion). By the way, this fact confirms once again that education is more important than genetics, although, of course, one thing does not interfere with the other.

9.586. The golden mean of evolution.

Evolution is not only one straight branch. This is a kind of bush with lateral, parallel and dead-end branches. These branches fight for their place under the sun, the weak ones die out, the strong ones go to a dead end and only the golden mean grows higher and higher.

9.688. Humanity evolution is evolution of food.

Humanity evolution is evolution of food.

10.571. There is no reason to think.

The paradoxical fact, with the development of science, people are becoming more and more dense. The more sophisticated the predator's weapons, the less brains it needs.

10.572. The main motive.

Perhaps, the woman is the main motive of development of human civilization. For most men, lust and women are the only motive for which they are able to think and work.

10.655. The evolution of the void.

There are two kinds of emptiness, from one you move away, to the other you approach. But wherever you go, you come from where you came from.

3.1398.

The smooth evolution of the five turns it into a six. Honors, finished school, become six, if you temper pride.

3.1534.

The question of evolution is about the survival of the virus in the environment, the immune system is persistently struggling with it.

3.1582.

Sweet and began to ferment fruit called the monkeys of the species homosapiens a terrible break-up and procrastination. To not die from laziness and hunger, Homo had to start hard to think. To be able to indulge in their dreams and pleasures, the monkey was forced to show great effort and intelligence.

3.1592.

Homo monkeys have developed intelligence through a combination of the following factors. Excess sugars in food. Access to salt and the ability to eat salt in large quantities. The third factor is the loss of its ecological niche. The fourth factor was meditation on fire. The fifth factor is hands, jumps and grasping reflexes.

3.1736. What about the monkey?

Freedom is very different. There is freedom and there is freedom. ...and there is still such freedom and such, and even such freedom. These are all very different freedoms, often mutually exclusive and hating each other, ...depriving each other of their freedom, killing each other... Freedom from Vice and blemish free. Freedom from the worm and freedom of the worm to devour you. A particular challenge is the simultaneity of freedom. For example, as crosses the freedom of the pig to eat an Apple seed - grow Apple-tree - to breed the worm and Apple... and most importantly, how do you relate to all this?

3.2404. Evolution from point to point.

Because expansion is narrowing, expanding, sooner or later you will become a point.

3.2901.

Happiness is incompatible with the progress of civilization, happiness makes a person lazy and sleepy. But happiness as a drug, as a mask of suffering and a source of narcotic withdrawal is quite suitable for the purposes of the progress of civilization.

4.1119.

Two-dimensional people are people frozen on the second step of evolution. But there are those who are stuck on the first.

4.1241.

You are not restrained in love, your embrace is like a hot cauldron, like a monkey fire.

4.1902.

There is no tree of life, there is a kind of canvas, a continuous intertwined network where everything is intertwined with everything. Interestingly, at the cellular level, all organisms are constantly changing genes. Viruses carry bits of gene structures from one species to another. The evolution of one species produces chain changes in other species living next to them.

4.2187.

They are no longer animals, but they are not gods yet. An intermediary stage of evolution? Some semi-material. They call themselves humans but that is false. The only human I know is God.

4.2192.

Animals cannot act differently from what their instincts demand from them. Humans can. Man has a free will. I want to be a human very much. I read a lot about real humans. I am not personally acquainted with any real human. They are said to have been extinct long ago or, rather, evolution has not created them yet.

4.2528. Hunters catch monkeys on greed.

You cannot take the carrot, avoiding the stick. If you like gingerbread, put up with the whip. If you don't like the whip, give up the gingerbread.

4.3424.

The idea that if you relax, people will definitely attack you is greatly exaggerated. People are hardly predators, rather, they are cowardly monkeys, whose brain pulsates with only one thought: to run away or not to run away.

4.3648.

The evolution of life is monstrously slow, but perhaps this is for the best? They say you only need to hurry when catching fleas.

4.3904. A garden of Eden, vain butterflies and hungry caterpillars.

Modern people are no different from the ancients. Spectators demand bread and circuses. Actors are full of vanity and demand power. Civilization actors and their viewers. Civilization gods and bots. Bots need bread (that's money) and pleasure (the spectacle). The gods need admiration, they want power. On the other hand, are they gods? Who is the liar who called the actor God? The actor is a butterfly, a simple evolution of a caterpillar bot. God is more like a garden. God likes butterflies, but flowers and birds are even more joyful to his heart. The latter he feeds fat caterpillars.

5.1867.

According to Darwin's theory, man evolves to overcome difficulties in his environment. If problems, to evolve, man ceases and he begins a neurosis. The lack of evolution causes terrible human suffering. To get rid of suffering, a person must invent problems for himself. You can start with stupidity, you can dream. How you make up your own problems doesn't matter.

5.1980.

Before the invention of fire, monkeys mostly ate sweet fruits, which did not contribute to the development of intelligence in them. Excess sugar is bad for socialization and the ability to focus. By reducing the consumption of sugar, the monkey became calmer and was able to focus attention, that is, learned to love and think. Concentration is love.

5.2035.

Man did not immediately evolve from a monkey to the crown of evolution, each of his upgrades cost him millions of years and his own reasons and circumstances. So you train each skill separately.

5.2104.

Macaque is a transitional stage of evolution from animal to God.

5.2118.

People don't lie, lie only monkeys, for which he suffered.

5.2119.

Servants of the Gods help people, but not monkeys.

5.2388. A dead-end branch of evolution.

The dead-end branch of evolution is not a mistake or a problem of evolution, but a branch of the tree of evolution. The tree needed branches, they grow fruits, flowers and leaves.

5.2392.

The older the individual, the more dangerous her breed, with age, the number of mutations increases. Reproduction of old individuals will litter the genetics of the nation and lead to degradation and extinction of the species.

5.2514. Cowardly primacy.

The eternal fear of falling into the abyss haunts any Primate from birth to death. What if you jump into an abyss and there's no branch at the other end you can grab?

5.2530.

Human civilization was born of sex between a man and a woman. In fact, sex created monogamous love and affection between male and female. This technical solution reduced competition between males and allowed to form larger groups than in all other animals.

5.2633.

The absence of hair on a man tells about his more civilized and less aggressive. The homosapiens lived in large crowded groups, full of parasites, wool in such conditions was unnecessary. The larger the group, the less aggressive people are, more intelligent and less wool on them.

5.2755.

Angels and demons are just birds and rodents that are as far from God as all other animals – mice, rats, fish, insects and monkeys. By the way monkeys are particularly fond of bird eggs and did not disdain even roasted snakes.

5.3058.

Human greed is everything from monkeys, why people tends to rise, because he is a monkey and wants to climb the tree.

5.3088.

He who does his work without love is like a monkey who imitates everything he likes ...getting carried away with this and that.

5.3157. On a branch of a tree.

I've noticed that a normal person always lives on the brink of madness. A man walks on the edge of madness all the time. If a person stumbles and falls, he will become crazy, and he will either plunge into the abyss of pain and fear, or fall into the euphoria of a balloon. But there are also those we call geniuses, those who grow wings. Geniuses are like angels, they can fly in darkness and in the light of day, looking for energy and information, and when they find it, they can return to their nest, to their thin branch on the edge of the abyss of madness. Monkeys have always been jealous of birds and dreamed of learning to fly. Geniuses are those who managed to do it.

5.3181. Wooden idol.

Initially the monkey prayed for the tree, because the tree fed it. Then a lightning struck the tree and it burned down. Thus the monkey realized that the real God was this fire. However, many idolaters still pray for the trees, so we call them wooden and oak-headed fools. Smart people are those who pray for love, that is, for fire. However, those who are enlightened know more than anyone else.

5.3182.

Monkeys have always had at least four idols... The most cowardly prayed to the bear, for he was terribly afraid. The wood was praying at the tree. The weakest, who were driven away from the tree, began to pray for the cow and the bull, for they saved them from hunger. Well, the most visionary began to pray to the fire, for they noticed that love and passion are the strongest.

5.3307.

In every man lives a fierce beast, the task of the mind-to subdue him... Accordingly, those who are weak in mind, unable to control their inner monkey and therefore we call them macaques.

5.3767.

The trend towards unisex and hermaphrodism is not progress, but degradation of evolutionary forms associated with the simplification of brain functions. Primitive animals were originally hermaphrodites, but evolution and the need for specialization separated them. On the other hand, the perfect also have hermaphrodism because of the lack of need to save and concentrate energy. Therefore, the development of society will give rise to hermaphrodism in both the lower and the higher. In other words, Homo and bi sexuality are characteristic of both the strongest and smartest and the dullest and weakest.

5.3949.

Man is a monkey who thinks he is an angel, but that does not spoil him, but on the contrary, for the life of angels is much more pleasant than a macaque.

5.3950.

Of course, the monkey is vain and thinks himself an angel, but it is this vanity that turns him into a man.

5.3957.

Lyrically speaking, do what you want. Monkeys do what they want. People and Gods do what is necessary and should be done.

5.5436.

Evolution is an endless improvement, the one who does not want to improve turns into a fossil and dies. Oil is the blood of the dead, coal and stones of their mortal bodies. Death will fuel life.

6.1349.

The essence of reason is that the mind can set a goal and to achieve it begin to create tools and means of production. The mind with one instrument creates a second instrument, a second a third, a third a fourth, and a fourth achieves the goal.

First you create your mind, then your mind begins to improve and train your body and yourself. Further, with the help of the body and the power of the spirit, the mind creates other means of production by which it earns money and achieves its goals.

6.2725.

Man is not a body, but a mind. The body is just a pet monkey.

6.2885.

If you live in a society, then you have to serve that society, otherwise leave... But you can't live without society. Society creates man, without society you would be a miserable monkey on a palm tree.

6.3332. Rattles and mirrors.

Your body is quite an expensive thing, so there is no greater joy to the devil than to buy it for a song. Beads, toys, candy, mirrors... Stupid monkey is ready to give everything for nothing worthless things and pleasures.

6.4076.

Man is made of mud, that is, earth (minerals) and water.

6.5022. Evolution from point to squirrel.

Learning and self-learning, in particular reading, generate the development and growth of the individual, thereby expanding the range of tasks that this person is able to solve.

6.5031.

The human " I "consists of two parts: the first is the animal" I", which contains the instincts and needs of the animal, and the second is the inner voice generated by speech, books and communication with other people. The inner voice is like an outside parasite lodged in the head of a humanoid monkey, but this parasite is the human mind, and that other " I " is just an animal (or maybe a God).

6.5032.

The essence of smiles and flat muzzle anthropoid apes in that it gives rise to the mind. Neanderthals had a longer muzzle, which prevented them from speaking and, therefore, they were less intelligent. The ability to smile signaled the possibility of communication and conversation, and therefore was the Foundation of intelligence.

6.5135.

The human mind is an information virus, a parasite that has captured the brain and body of an eternally unhappy monkey.

6.5291. The mind creates tools.

The essence of reason is not that the monkey took a stick and used it to get a banana. And that clever monkey said stupid to use the stick to get a banana to take third, and the rest to give to her.

6.5466.

You want to beat the monkey, distract her attention with a banana.

6.5793. Beauty will save the world by uniting it.

The highest goal of love is to achieve unity. The most important goal of the development of human civilization is to achieve the unity of mankind, the creation of the Confederation of planet Earth.

7.1035. Not Darwin.

There is no evolution in the direct sense of the word. A set of genes is defined with great redundancy and as external circumstances change, certain genes are turned on or off. However, you can see that the genome itself develops within itself, regardless of the rest of the organism, apparently, this is the real evolution.

7.1038. From the theory of species evolution.

The more people are alike, the more they want to be different from each other.

7.1046. Evolution Of God.

It was thought, but God is also evolving. If the whole world is God, and we see that the universe is expanding, nature and man are developing ... then it turns out that our God is growing and becoming stronger and stronger.

7.1617. Throwback of evolution.

The most important thing for most people is to eat and mate. This is due to the fact that the last million years, these two desires were a guarantee of survival. By the way, that's why seducing a girl without taking her to a restaurant before, bad form.

7.2429.

WISDOM IS RESTRAINT. Development of the human brain can be reduced to evolution of the frontal lobes, which are in charge for the chaos of desires and thoughts living in the head of a wild human. The better a person controls his desires and emotions, the more sensible he is.

7.2483.

Lie is subjective reality, truth is objective reality. The mind lives in a subjective reality, so lying is its natural habitat. The essence of personal growth is the evolution from lies to truth.

7.2820.

Evolution goes on in waves – rational people gradually evolve into stupid people.

7.3353.

When a tree falls, the monkeys run away. It's a proverb about idols. Leave the sheep alone, attack the shepherd.

7.3363.

They say the pursuit of perfection is the evolution of lies into truth. It is said that when lies become truth, that is perfection... they say perfection is the end

7.3501. Question evolution.

The first thing to learn in life is not to take anything to heart. Your pain point is the Achilles' heel, where enemies will beat and beat until they break you. Nature does not like vulnerabilities, so it will beat them until the weak individual dies out.

7.3523.

Natural selection is the basis of life and the main mechanism for the evolution of living beings.

7.3525. The law of exceptions.

Universal biological law, according to which, excluding lies, nature leaves the truth. Natural selection and evolution work on the basis of this law.

7.3533. Evolution of perfection.

The evolution lie is a movement from one extreme to the "Golden mean" and the attainment of perfect truth. The evolution of truth is the evolution from point to ball by absorbing all its extremes.

7.3555. Question evolution.

They say God can't be wrong, it's not, and evolution proves it. However, God quickly corrects his mistakes, quickly erasing them from the face of the Earth.

God the Creator creates a new infinite-best, erasing the worst of the old.

7.3752.

Trying not to think about the white monkey turns it from an illusion into reality.

7.3895.

Once human noticed the power of fire and wanted to serve it, that's what made him sensible. Hope and desire to negotiate with fire, tame it, share its power created Human.

7.3896. The human mind created the cold.

Once, when man was still an animal and he was cold, he noticed that lightning from the sky hit a tree and fire appeared. Fire gave warmth and saved from the cold. So people noticed that the fire eats the tree and realized that the fire gave heat and he needed to serve and feed his tree. So the man opened the fire. To cause fire and communicate with God, man learned to use flint flint and so learned to make stone weapons. Then the man noticed that the fire consumes not only trees but also animals, and realized that the fire needs to feed and meat. So people started hunting, roasts and learned to eat grilled food. Stone tools helped man to hunt and work with wood, taught to build dwellings, to process skins and dress. Feed the fire flesh was very hard, had to learn to hunt, conjuring up weapons, develop in itself the mind. Not all could to hunt, some decided feed the fire plants. So there was first gathering, culinary arts, medicine, and then agriculture.

Believing in his God, man began to think about his nature, about how to serve him, about how to propitiate him and what other sacrifices can be made to him, so that he gave warmth and benefit, and not pain and fire. In observing and meditating on God, man learned to think and created philosophy. Trying to communicate with God, he created drawing and writing. Trying to combine fire and earth, have learned to extract and process metals, ceramics created. In thinking about God came to the concept of unity of earth, water, air and sky. Created funeral rituals and eventually moved to a sedentary lifestyle. He created the concept of grain, learned to grow grain and make bread, created agriculture. The subsequent history you know.

The symbols of God are fire, light, warmth, joy, sun, grain, intelligence, love, hope, faith, energy.

7.5251.

The evolution of the games will lead to their merger into one mega game, which will be called the universe. Current version of the Universe is v.3.0

7.5350.

Even if people evolved from monkeys, it doesn't change the fact that god created everything, including monkeys. Moreover, the idea of evolution corresponds to the religious idea of a grain that should grow.

7.5359.

Either stop this degradation. Time does not stand still. From time to time, you degrade. Evolution is a movement in step with the times.

7.5360. Homo Religiosus.

It is scientifically proven that man was made to the image and similarity of monkeys, but science has also proven that human intelligence is rather of divine nature.

7.5361. Working monkey.

Neanderthals already had well-developed areas of the brain responsible for understanding speech, while the speech apparatus itself was not developed. Question: who were they listening to so attentively?

7.5370. Blind branch of evolution.

Equality in marriage is welcomed. And though a donkey and a horse may have offspring, mule won't have it anymore.

7.5371.

Evolution is said to be a gradual development, but evolution also leads to abrupt revolutionary processes when a critical mass of changes is reached.

7.5716. Just don't look.

Eat often and little by little it is a natural type of food human primates over millions of years. Hungry, ate a banana and went about their business, and so as the feeling of hunger. However, it is important to remember that this works in a warm climate, and in a cold climate animals accumulate fat in the summer to starve in the winter. In the winter, as much food, the animal will attempt to eat everything he sees. So in winter, the less food you see, the better.

7.6126. A broken branch.

Experience suggests that the monkey alone could not get off the palm tree, it clearly helped the case.

7.6448. God Of Fire.

Human civilization came about through meat. God taught man to eat meat. Man, seeing God eat meat, and wishing to join the power of God, shared a meal with him. Man cannot eat raw meat, but roasted meat, the food of the gods, was to his taste and gave him strength.

7.6582.

There is no free will... in the soul of man lives God, the snake and the monkey.

7.7162. Adam, contemplating fire.

They say the first man was Adam. It is likely that it was. Man is mind, the first monkey to receive revelation from the contemplation of fire, was called Adam. Then Adam found a wife and her name was eve. Eve admired and believed in her husband, and took from him his vision of the world. The mind gave Adam greater force and power, he became head of tribe and priest. The tribe quickly gained power and subdued other tribes. This couple had many children, which parents have got power, the beginnings of perfection and reason. Mind gave these people power and they became the dominant species, gradually began to control the whole world.

7.7164. The Fall Of Adam.

Becoming the head of the tribe of monkeys and gaining great power and power, Adam fell into pride and a Legion of other vices. They say he came to a bad end, but history is silent on the details.

7.7572.

Nature, caring about the evolution of the human spirit, takes care that a strong spirit rose, and weak in spirit faded into oblivion. Therefore, the weak in spirit cannot resist vices and sins... and, having fallen their victims, fall into hell.

8.1134. Slow evolutionary process.

The only way to avoid degradation during growth is to adhere to the harshest harmony in evolution. Any unevenness, distortion or excessive acceleration in growth breaks the energy balance in a person and causes degradation of other vital systems.

8.1199.

Don't criticize women for aggression, feminism, impudence, love of freedom and other stuff like that as all those qualities of female soul were given by nature so that every woman would wait for the right man whose reproduction fits the necessities of human evolution. In other words, it's necessary for every woman to find true love and any mistakes about it are really unfavorable.

8.1200.

Feminism is not a matter of female self-identity, but rather a matter of the mind degradation of males. There are very few men who deserve love and it's really sad.

8.1590. Direct malicious intent.

Indeed, human beings were originally herbivorous animals but evolution made them carnivorous. Proper food nicely influenced those animals' mental abilities and turned them from apes to humans. But some people's attempts to reverse the evolution and make humans become herbivorous apes again evoke disputable arguments.

8.1619.

Visual patterns (including 3d) is kind of a parallel information port to the human brain. And considering that the eyes are the most ancient way of transmitting information, obviously it's the most understandable for humans. But there's one thing. Sequential method of perceiving information through reading and words is what turned apes into humans. The mechanism of interpretation and transformation of sequential code into visual conceptual one, is in its essence, a half of what human intellect really is. That's why learning things by using video sources and neglecting reading makes people dumber.

8.1708. Pencil.

The task of evolution is to make a sharp person out of a blunt one. Person reminds me of a pencil that should be sharpened for the good of the cause.

There's nothing sadder than a pencil lying in a drawer without being used. Created to draw, it turns into pain and sadness without work.

8.1709.

The task of evolution is to make a person-homoreasonablus out of a person-homofoolus.

A homousefulus out of a homouselessus.

A homostrongus out of a homohelplessnus and homostrengthlessus.

Overall, at the moment we are seeing a transition period of human evolution from some primitive forms of life into homo sapiens. At the moment, no more than 20% of the human population has evolved into homo sapiens. There is still a lot of work ahead.

8.1719. Imperfection of evolution.

If a person is impenetrably foolish, it means that he is a still alive Neanderthal, stuck in the transitional stage of evolution.

8.1723.

Many people are very similar, the variety of models leaves much to be desired. I dare to suggest that once they were completely the same, now it is less common, because evolution does its job.

8.1748.

Women are the evolution driver, choosing the best men, they move the civilization forward.

8.1889.

Humans are quite wild animals, a fine cover of civilization ten thousand years thick barely hides the millions of years of his true nature.

8.2618. The danger of knowledge is wrong interpretation.

8.2655. Evolution into man.

To kill a slave is difficult, but squeeze it out gradually ...on a daily basis... just a little.

Squeezing out a slave is a long and painstaking process, we can say, evolutionary. You can't just kill a slave inside you, because then you'll be an empty trash can... It is necessary in parallel, as the extrusion of the slave, to grow a man free and noble.

8.2917. Scourge of God.

Mankind is the cause of many species mass extinction.

I've just thought. There are many versions of the extinction of dinosaurs and another 90% of species in the Cretaceous-Paleogene period 65 million years ago. Let us suppose another one: there appeared some kind of intelligent life, which eventually killed everyone. And an important consequence of the intelligent life's influence on evolution is a decrease in the size of animals. People try to kill large animals first and small ones are more likely to go unnoticed and survive.

8.3113. In matters of the nature of things violation, the nature cannot be trusted.

The matters of survival of the weak and the health maintenance of the useless and the old contradict the ideas of evolution and renewal, therefore in these matters the nature is not an assistant to human, but his implacable enemy.

On the other hand, the nature cultivates strong, useful and beautiful individuals... Be in trend and the nature will help you.

8.3333. The key is not to overdo it.

Labor created human from a monkey. But labor also created some insects. For example, ants sleep 4 hours a day by fits and starts. Bees are buzzing constantly, too.

8.3414. Thought is the greatest heritage of humanity.

Human thought is the greatest achievement of evolution.

8.3618. Without money to live cannot be, as cannot be to live without food and clothing.

Well, unless you're a monkey on a palm tree.

8.3889. Devolution.

Monkeys don't want to turn into human as they are too lazy to work. If a monkey started to work, it would turn into human at once and the opposite is respectively true.

8.3892.

Evolution is a complex step-by-step process. Upgrading skills one by one tediously for a long time, it`s possible to gradually become human.

8.3893. World of primitive gods...

It`s occurred to me that life in the virtual worlds doesn't' include work. It's known that without work human devolves back to animal. Could it happen that virtual worlds are full of wild super-people, devolved back into monkeys.

8.3914. Humans are the only imperfect animals.

Humans are imperfect. Their own imperfection causes them great pain and humans become much better and more perfect by running away from this pain. Humans are the only animals that feel the need for perfection.

8.4005.

It can be assumed that human isn't the end product of evolution at all, he is more likely to be some kind of error... System failure, some kind of virus that evolves and grows so fast that the immune system of the Earth hasn`t reacted yet.

Probably, there is some virus which, getting into human brain, made it work some unprecedented way.

8.4025. Theory of evolution..

Not all people participate in the evolution, many skip it.

8.4146.

There`s a view that various parasites highly influence evolution, human in particular. All these mosquitoes and other things, by transferring foreign DNAs, including virus ones, from human to human (from animals to human.), highly influence his inner world...

8.4150.

Education is not very beneficial to the State. Sensible people are stronger and freer than those steeped in obscurantism and absurdities. It is difficult to control free and strong people, since their brain filters cut off the propaganda and absurdities being spread by the media that washes the brains of anthropoid apes.

8.4382. We need surplus for exchange.

Superfluous created civilization. Human civilization is the civilization of superfluous things. Surplus is necessary for exchange, it created trade. If everybody had only what is necessary, there would be nothing to exchange and civilization would die along with merchants, workers, peasants and even women would lose the reason to stand men.

If nobody can give anything to anyone, nobody needs anyone.

8.4928.

Suffering turned the monkey into a man. Suffering from cold and hunger, she climbed down from the palm tree and was forced to start working to escape from her suffering.

8.5525. Reasonable philosophy is the driving force of evolution.

They say evolution is a slow process, nevertheless I´ve seen the people that evolved from insects to rather sensible primates over 7-8 years just by embracing reasonable philosophy.

8.5531. The incredible variety of Beauty.

By cultivating imperfection, civilization contributes to the emergence of various types of perfection. In the past, natural selection used to cultivate only a few types of perfection and the rest perished... However, now that the weak have the opportunity to survive, the burning issue for it is the search for available niches to achieve perfection.

8.5554.

Man is a new type of monkey. Monkey, streamlined for life in the team.

8.5598.

Problems and difficulties are good for the population... Natural selection improves evolution... The worse things get, the better people.

8.6020. People boring.

It is boredom that distinguishes man from APE ...A person is able to concentrate and do the same thing for many years.

8.6031.

The monkey did not just descend from the palm tree to the ground, it fell quite hard and, breaking half of its bones, was unable to climb back...

8.6033.

Reason is rather a by-product of pleasure-hunting... A candy is a good reason to reflect upon it.

8.6119.

It's easy to be smart when you are all smart and special, but try to turn your brain off being just a typical sheep. Evolution of a sheep into a human, that's what is really valuable.

8.6204.

Curiosity is an important motive of evolution.

8.6340.

Evolution engenders the diversity of species, for example, such humanoid species as как reptilians, zombies, «flower children», «animal spirits and so on will be set apart from people over time.

8.6366.

On the example of "Variothoughts" you can trace the evolution and development of thought.

8.6367. Evolution of philosophy.

Man is his thoughts and philosophy. They say that man does not change, and it is true, but his thoughts change... his philosophy changes... in the last forty thousand years, man has not changed much, but his philosophy has undergone tremendous changes.

8.6369.

The evolution of human civilization is the evolution of philosophy.

8.6385.

Even though civilization hasn't contributed to human evolution, it led to the evolution of philosophy. People haven't changed, but their philosophy has become more perfect and stronger.

8.6386.

Civilization is the evolution of thought, the evolution of philosophy.

8.6388.

Civilization hasn't resulted in a significant evolution of hardware, but it has caused a major evolution of software. That is, the human mind has remained almost unchanged, while his soul and philosophy (operating system) have significantly evolved.

8.6463.

The analysis should be done deep into at least two or three steps, for example, all people are descended from Adam and eve, who had three sons... In the second step, we have to agree with Darwin that these three didn't have very many options...

8.6487. A greenhouse of decent people.

Stronger we are many who wants to do it, and happier - one, is because this world is not a zoo where the monkeys are giving out free bananas, so they were happy and entertained the audience. The world was created to raise perfect people.

8.6659. Idiocracy.

By defeating evolution, civilization allowed idiots to survive and flourish, turning them into the primary form of life.

8.6830.

Humanity needs upgrade of philosophy, otherwise man-made technology will become better than the man himself, due to science development. Evolutionally, this situation will lead to degradation and dying out of human as a non-competitive species. Human is so sinful and vicious that nothing will prevent him from self-destruction over technologic progress.

Technology will turn human into God, But being God is not as easy as it seems. God needs special philosophy, special understanding of life, otherwise one wrong thought and the whole world will perish.

8.6967.

Without a kick in the ass monkey with palm trees did not want to get off.

8.7178. Evolution.

Scammers, fraudsters and charlatans evolutionary mutate much faster than their victims. Idiots, too, evolve, today's idiot on order smarter idiot half a century ago...

8.7419. A flaw in evolution.

A decent man on the outside and a macaque on the inside.
Many macaques very successfully disguise themselves as a man, until you talk and do not understand who it is.

8.7526. Good monkey.

He spends a hundred, earns a thousand. It Manager-knopochek, the slot machines, in which he throws coins and presses the button... And they dance ...they ...they sing ... they work.
Each pet monkey brings its owner to spend a coin seven coins profit.

8.7592. A double-edged weapon of nuclear destruction with a nuclear warhead.

Love is the most powerful weapon in this universe, but, as any weapon, it is deadly in the wrong hands. Empty brains turn love into poison. A monkey in love is a horrible man.

8.7685. Technological progress will make possible classless society.

Basically, I agree that society's evolution seeks to achieve classless society. Society with no cities or villages, no physical or intellectual work, no slaves or oppressors. Frankly, I still can't imagine society with no idols, though.

8.7861.

The stronger, smarter and more independent a woman is, the more difficult it is for her to find a man who would inspire her into reproduction.

On the other hand, such state of things speeds up evolution and gives the best men a huge priority for reproduction, while the worst ones die out. Now, when a woman is independent and able to feed her offspring on her own, the best males are able to have tens of children born by different women.

8.7863. A monkey with a grenade.

The problem is that people fail to keep pace with technology. Human adaptive mechanisms are intended for long, natural factors when it takes 50 to 70 generations to adapt. However, now the fact that a civilizational leap has happened in 200 years and that people have hardly changed gives rise to a problem similar to the monkey with a grenade.

8.7930. Evolution of a worker into an artist.

Although our society is post-industrial, it does not mean that people should be left without jobs or that they should work in the service industry as salesmen, waiters, etc. It rather means the transition to the production of unique, non-mass market goods, made with a high degree of novelty and manual production. It is moving away from cheap mass production towards individual, piece production with a large surplus value. It is the moving away from handicraft and stamping towards art. Production is not a craft, but an art in the post-industrial society. Now, it`s not a craftsman, but an artist who should be engaged in production.

8.7945.

Millions of years ago, a monkey climbed off a palm tree and could no longer reach bananas, so it had to learn to use a stick.

8.7959. Computers evolving into humans.

I see a future in which every home computer will be an individual, independent personality and the communication node of artificial intelligence... That's not all: human intelligence will be next to artificial intelligence... Will they come into conflict? Will they form a symbiotic organism? Or maybe one of them will prevail and will consume the other.

8.8094. Space does not tolerate emptiness.

Emptiness in the mind cause pain and aggression in anthropoid apes.

8.8142.

Brain evolution is wavy... He is improving, then degrades, and this phenomenon can be traced even within one person in the course of his life...

8.8306.

The first of those who find a free niche, entirely fill it and don't let others in... Others will have to look for a new niche with energy in order to become first in it.

8.8415. People-fungi.

As the science of paleontology shows, humans are descended not from Eve and Adam and not even from apes, most likely, humans are descended from algae... Algae, mosses, lichens - these are our real ancestors.

8.8537.

You need strong enemies, because to survive, you have to be smarter and stronger than their enemies. A good enemy is your ladder to the top of evolution.

8.8545.

Kindness is one of the cornerstones of the mind's evolution. Animals are very aggressive and rarely help each other. Kindness is a mode that favors exchange of information.

8.8546. Love is the basis of Reason.

Love created human reason. Having learned to love, humankind became Kind. If you're kind, it's easier for you to communicate with others, exchange and accumulate information and resources. Love allowed humans to start communicating, which provided a basis for reason and civilization. Animals don't know what love and kindness are, but humans do. Love may be said to have created all other human emotions too. We do remember how close hatred, vanity, envy and so on are to love.

8.8598.

Supposedly, the human soul is a kind of virus or parasite that once invaded the monkey's body. Major contradictions between aspirations of the soul and those of the body support this fact, as well as two completely independent reproduction systems. DNA is responsible for the body and external things, such as social factors, books and other external sources of information, are responsible for the human soul. Specifically, the human soul is an essence structured from outside rather than from within. Moreover, if you open up the cranium, you'll see that a barrier separates the brain from the rest of the body and is in immune conflict with it.

8.8617.

Processes relating to immigration and overpopulation created homo sapiens and civilization in general. Forests became savannas and monkeys climbed down palm trees and went over the hills and far away in search of food, gradually populating the planet. In this regard, I think any restrictions placed on migration processes are likely to produce counterproductive effects and to lead to human degeneration.

8.8618. Edge mind.

Humans have always lived at the edge. Primates moving between trees used to balance at the very edge of branches. When at the edge, you always have to jump into the precipice: it's your only chance to jump over to the next branch without hurting yourself. Human beings are marginal, edge creatures always living on the edge and their lives are a delayed jump over their heads associated with constant risk.

By the way, apples also contributed to the emergence of the human brain. Humans used to eat fruit and had to remember their color, maturation time and place, all of which required the development of memory.

8.8636.

Undoubtedly, human bodies evolved from monkeys, but the human spirit, his soul, collective mind and society could have well descended from gods or extraterrestrials, under certain circumstances.

8.8653.

Sectarians like to promote their followers the idea of vegetarianism, this is due to the fact that the lack of energy in the body turns off the brain first of all and such people are easy to manage. And really, people were originally monkeys-fruit-eaters, but it is meat that gave them extra energy to pump their brains. Being left without this energy, human turns back to who it used to be before the era of meat.

8.8654.

We can say that the rise of human civilization over the past hundreds of years is associated primarily with the ability to eat well and eat meat. Having received additional energy, human got a massive opportunity to think.

8.8662.

Humans have evolved into an intelligent life form because they needed to, whereas those who did not are still monkeys.

8.8727.

As rightly observed by the Hindus, this world is ruled by elephants, tigers and monkeys. Many gods have many faces and gods with woman faces are extremely violent.

8.8869.

If the monkey was all right, he would never get off the palm tree.

9.1262. Maybe everything was the opposite.

Let's suppose that the human brain is a vivid example of symbiosis, when some parasitic organism, let's call it "the It", captured the brain of a mammalian monkey, thus provoking accelerated evolution of its mind, as an answer and counteraction to this parasite. Attempting to gain freedom and get rid of the parasite, the mind began to develop, gradually reaching the heights of the intellect and the fortitude.

9.1314.

From the point of view of evolution, it makes sense to reduce the level of emotionality in the organism of a modern human.

9.1591.

Human got on his feet, as he seemed bigger this way, and many predators started to be afraid of him and avoid him. The one who is bigger, is stronger. Human, crawling or running on his hands and knees has minimum chances of survival.

9.2728. Price of a spiral.

All human problems are a consequence of the violation of rules, but the very opportunity to violate the rules is provided by the system and also occurs according to certain rules and is accompanied by problems and suffering. Why is this done this way? - The system becomes more stable, flexible and capable of evolution this way.

9.3290. The evolution of the universe.

The universe evolves from a unidimensional to a multidimensional space by storing information. The exponential growth of information leads to a collapse of the mass under its own weight and to the system's production of critical mass and a supernova explosion. Specifically, there are two explosions: one comes outside and the other inside itself. The explosion inside itself produces a breach into Nothingness, a superconducting point, on the other end of which a new universe emerges, of greater dimensionality than the current one. Energy unpacks and flows from this three-dimensional world into a new seven-dimensional universe, thus creating it. The supernova gradually works off its energy in this world, becomes depleted and dies out, giving less and less energy to the new world. When energy is up, the new world, bereft of its source of energy, will collapse under its own weight and, folding back into a point, will produce critical mass, and everything will start over again.

9.3657. The evolution of man into an angel.

The path of perfection is the path from grain to wood.

9.3790. Women block.

The choice of women-voters is dominant and critical for a democracy. Women are very attentive to what other women like and so they are inclined to vote as a unified block. Since the beginning of time, the power of leaders was based on the females` support in the monkey communities.

9.3839.

Civilization allows the weak to survive and weakness is the basis of mind. The weak, trying to achieve perfection at least in any-thing, reaches truly unique heights.

9.3932. Evolution.

It is too hard to do everything right, especially at the initial stage. So first you can do as it turns out and later, when you get stronger, you will evolve into more correct forms.

9.4329.

Success is a matter of evolution or revolution. Revolutions are dangerous and calling for courage or stupidity. Evolution is harmless but it demands great patience and faith into the invis-ible.

9.4332.

Beautiful is what you love. Reasons love can be many. The man looked closely. Monkey man. Man loves truth. A man... can inspire a lot of love.

9.4346. The ability to see.

Brain means the ability to see and use resources. As the evolution goes, humans gain the ability to derive use from more and more resources that were unavailable once. Perfection is the ability to derive energy.

9.4517. Monkey instinct.

Man can love the different, because he is good and kind, it is pos-sible for that strong and perfect... and you can, because it is loved by others...

9.4561. Grow an Angel inside yourself.

The meaning of human life is the evolution of a human into an angel. When a person attains perfection, this person turns into an angel.

9.4639.

The struggle for existence and vices have turned the monkey into a man. There will be no struggle for existence and vices, and civilization will disappear.

9.4992. The engine of evolution.

Smart and strong woman is the engine of evolution, because for reproduction she will have to find an even smarter man than herself. With stupid and weak men smart and strong women do not want to breed.

9.5041.

Women are the best hope of evolution. Women should prevent idiots from breeding.

9.5281. Evolution of worms.

The truth is, in reality, we're worms... and we live in J. But I really don't like this reality, and no matter what anyone says, I don't want to love what's there. The only thing we have is mind power. The mind, which possesses such tools as faith, hope and love, is capable of miracles, that is, it is able to make the unreal real.
In General, we evolve, from the easiest worm to evolve into a caterpillar, then a butterfly.
- But how does a butterfly get out of its ass?
"It doesn't matter. Butterfly is not the crown of evolution and not our goal. From a butterfly we evolve into a bird, then into a dinosaur, then there are options, but I think a macaque will be a good choice, and finally from a macaque we turn into a man...
- When do we get out of our ass?
Never!!! Here we will break the rules and go not down into the light, but up into the darkness, where the brains are... You need to take control of the brain.

9.5809.

The closest relatives of human are viruses, not monkeys.

9.5937.

The aim of the evolution is to make people different. Originally people were very similar to each other, but the cleverer they grew, the more differences between them emerged.

9.6597.

The first life emerged in geothermal springs even before oxygen, when sulfur was the basis of life. Metaphorically, it can be said that the first life emerged in Hell. Then it got into the ocean... And only a billion years later, during the oxygen revolution, it came out into the light, and seemingly, lives in Heaven since then.

9.6909.

I`ve noticed that evolution moves by leaps.

9.6928.

The meaning of human life is the evolution from zero to one. If you're already a one, you can continue to grow up to 3, 7, 21 and so on to space.

9.6992. Evolution of faces.

Their masks are similar to faces and faces to emoticons.

9.7449. Or at least not worse.

Running squirrels in a wheel is a creative dead end, the real evolution is a spiral. Spiral is when tomorrow is better than yesterday.

9.7450.

The spiral of evolution is a response to the changing facts of life. In case facts of life do not change, however, is evolution possible?

9.7580. Evolutionary growth is the preservation of good.

They say growth is when tomorrow is bigger than today. They say growth is evolution and increase. Usually, however, evolution is more about balance than growth. The situation is constantly changing in the direction of deterioration. Keeping good is evolution.

9.7609.

Robots are artificial insects. Robots are strong but their brains are as big as the insects'. Evolution, however, is not static and robots will soon become reptiles... At that stage, people are likely to die out because last time dinosaurs ruled over the world for as long as 60,000,000 years.

9.7628. The evolution of a point.

A point is unidimensional, a circle is two-dimensional and a spiral is three-dimensional. A hamster running round the system makes it dynamic.

9.7668. Viruses are cause of the mind emergence.

Human is an example of viruses evolution. Viruses laid the groundwork of the human mind, putting the idea of the endless self-destructive growth in it.

9.7669. Viral nature of mind.

Mind emerged as a virus, infecting the brain of the mammal primates.

9.7817. An old biblical story and yeast.

Meteorites brought intelligence to the Earth from space. Intelligence is a fungus parasitic upon the brain of mammals. Having come from space as spores, the fungus developed into mycelia and infected fruit trees. One day, a monkey overate apples and got infected too. You know what happened next.

9.8071.

The evolution of thought from zero to one is the development from the subjective to the objective, from an idea to its implementation into reality. The real world is objective: when you are working on a building or a car, there should be nothing indefinite or contingent in them.

9.8094.

By fighting emerging threats to itself, immunity is continuously educating itself and generating new conditioned reflexes in itself. In the same way, brains should confront and overcome a problem in order to acquire new knowledge. This will produce a conditioned reflex necessary to fight this kind of threats. The larger the number of threats confronted by brains, the better the latter's immunity becomes.

9.8186.

Evolution is always a reasonable compromise. The strong want to eat the weak, but the weak doesn't want to be eaten by anybody. Moreover, if the strong eats all the weak, the very next moment he will become the weak link himself and die of starvation.

9.8275.

The search for good spirits is the driving force of evolution.

9.8276.

Pain is the driving force of evolution: when fleeing from pain, a man runs to the future at breakneck speed.

9.8576. Infantilism is an unfinished program of growing up.

They want to slow down the maturation of their children. Earlier in 13 years they grew up and went to look for "their place in the sun", today they are still considered children. But what is 300 years of civilization versus a million years of evolution? This breaks their psyche, forcing to run away from reality in illusions.

9.8720.

The point of the evolution of a population is that a bad example sets a good one and now everybody knows what is not to be done.

9.9195. Little monkey.

One is dying to be like everybody else.. Everybody is on holiday... And I want it to. Everybody goes somewhere... And I want it, too. Everybody has a ring in the nose... And I want it, too. But it's not you who wants, it's your subconscious, the It that, being gregarious, adores to do the things everybody else does.

10.1382. Predators can not stand lies.

They say animals don't lie, they say lies are a product of the mind. As usual, everyone lies. Lying is a creation of nature, known as camouflage, weak animals pose as predators to be eaten less. The apes stood on their hind legs, shouting and waving their paws to make themselves look bigger and scare off the beasts of prey that were about to devour them.

10.1501. All for the best.

The human mind was created by the God - Fire, he tied it to warmth and pleasure. Roast meat tastes better raw. In the heat more comfortable than in the cold. Having begun to serve the fire, the monkeys became slaves of God.

10.1502. Reasonable pleasures.

After tasting the roasted meat, the monkey experienced great pleasure, after which it was hard to return to the raw food diet. From heat in the coldness, too, not very want. Had to monkeys from despair serve fire.

10.1719. Trapped in illusions.

When you become attached to unity, you fall through the earth. Trying to cling to the edge, you generate hardness and fall into the trap. You could let go of his hands and escape, but the monkey, grabbed a banana, can not release his hand, whether it prevents fear, or greed... it's hard to say.

10.1838.

Studying history and literature, as well as observing the present, I noticed that people over the past five millennia has not changed much... except that his stuff became stronger. This is a concern, a monkey with a grenade could be dangerous.

10.2301.

Mistakes and imperfections are the sources of innovation. Perfection is dead.

10.2437.

God is an idealist who has attained perfection and therefore he is never wrong. That's why he needs a man. The man is the one who is able to make mistakes and thus generate adventure, novelty and progress.

10.2469.

Monkeys are copycats. The more stupid and weak a monkey is, the more he imitates.

10.3676.

Uncertainty and randomness is a mistake. However, any living system must work with errors, because error is the basis of evolution, growth and life in General.

10.3677.

God is perfect, but mistakes and uncertainty are part of perfection. The system is designed so that errors do not kill it, but generate novelty.

10.3700. Divine impromptu.

As a humanist, I urge you to love mistakes, because mistakes are the secret of novelty and evolution. Error is life, while man is capable of mistakes, he's alive. Your craving for control is a craving for disease and death.

10.4546. The evolution of God.

Over I is an upgrade of I to my certain ideal state. The God of the new world takes power into his own hands to control IT through the Ego.

10.4972.

From plant food grows belly. Watch the herbivorous monkeys and be horrified. To be slim, the diet should be dominated by meat.

10.5950. Evolution of the universe.

God is the standard that has reached perfection. That means God stopped evolving and died. God is dead. It's sad, but ... he didn't die in the past, he died in the future. And this, of course, pleases. In the present, God is still imperfect and continues to create himself with his own hands. The perfect God, who died in the distant future, creates himself from the future of the past.

10.5988.

A lie is something that doesn't exist. Every fantasy and human thought is a lie. You are a lie, your mind is a lie. All your fantasies and plans are lies. All your memories are lies. The attempt to destroy falsehood destroys human reason, progress, art, and the evolution of forms. Death is the end of lies.

10.6009.

I have noticed that evolution does not always follow the most obvious path. The usual path of evolution is a strange and difficult path. The one who is difficult evolves, not the one who is easy.

10.6069.

The evolution of annelids into molluscs is a creative dead end, telling us the fate of those who are easy. It was harder for ordinary worms to live, so their evolution was more varied and significant.

10.6082.

Species appear and die out. Species occupy evolutionary niches, and species lose them. Species evolve and die out... The normal nature of things. This is called evolution. Are you against evolution? do you want to stop time?

10.6108.

If man, as a biological species, did not rise up and kill all the competitors, it might give the monkey a chance to become a decent person, go to College ...and all that. And so... well, bad luck to the monkey, what can you do?.. Even 50 thousand years ago, there were 7 types of people ...now there's only one left... And how

many of them before that was unknown at all. Homosapiens believe themselves to be gods and kill and eat everyone else ...and in
General... Homosapiens is the most vicious and monstrous predator in the history of the planet Earth. I think the monkey needs
sympathy, it just had bad luck with the neighbors.

10.6183.

Baboons on some spiral of evolution far outstripped man, which
killed them. Man with difficulty, but continued to evolve, and
the happy baboons went the easy way.

10.6415.

To do well, first you need to do bad, because " good "is the evolution of"bad".

10.6478.

Love the result and love the purpose it means to love, the end of
sex. Normal people love sex, not the fact that it ends.

10.6479.

You should love running, not the result. Loving running is like
loving sex, and loving the result is like being glad that sex is over.
When you run, enjoy the process, and do not think about the fin-
ish.

10.6495.

If there are no problems, you need to come up with them. Without problems, life is boring and there is no reason to grow.

10.6515.

To realize that man is descended from a worm, look carefully at
the sperm. As soon as the worm got its brain, it turned into a
human. The Union of sperm and egg is a metaphor for the Union
of practical experience and theoretical truth.

10.6552. Love is admiration.

You turn into someone you love... You take in everything that you admire.

10.6562.

You turn into something you love or get something you admire.

10.6583. Humility and pride.

The difference between a human society and a mycelium, hive, or anthill is the comparative isolation and independence of individual individuals. The hive is egoistic and full of pride, which creates in it fear and isolation from the world. People are more open and this gives them certain perspectives.

10.6584.

Animal this is a hive, not a bee, a mycelium, not a mushroom. So in the human community, the subject is the society, and the people it's just mushrooms and bees.

10.6596.

Potential is no better or worse than reality, it is just another form of reality. The reality tends to accumulation and the evolution of forms, nothing more.

10.6615.

Evolution constantly comes to a standstill, but does not give up...

10.6624.

Human society is something like a form of evolutionary development of mycelium.

10.6633. Trinity.

The very first living cell contained chlorophyll, mitochondria, and flagellum.

10.6647.

The living genome consists largely of the genes of various viruses.

10.6673. A single-celled fungus.

From too good life, mushrooms completely relax, degrade and turn into yeast. Yeast is very fond of sweet and rot. Yeast live where they do not have to take any efforts.

10.6678. The law is common to all.

Nature is very persistent. Evolution has tried to create a living cell many times, and it only succeeded on the 6th or 7th attempt. All previous attempts ended in extinction. The point is that if you don't pass step One, you can't go to step Two.

10.6680. A successful symbiote.

Lichen is one of the most successful slave owners in the world. He has enslaved a lot of cyanobacteria or algae, and forces them to release energy from the sun while he lies in the sun for thousands of years. Lichens live for thousands of years and do not know grief.

10.6683.

The mechanics of evolution is the process of exchanging information. The exchange of information takes place in the process of creating symbiotic unions in order to more effectively survive, defend and extract energy in a changing competitive and aggressive environment.

10.6687. The Trinity.

The ocean is a symbiotic organism, the unity of all living organisms that fill it. However, this is not all. The atmosphere is the same ocean. But this is not all. Underwater, above-water and underground combine into one single symbiotic organism.

10.6688.

The complex consists of the simple. DNA of the fungus and human DNA is similar to 55%. This means that, having created the mushroom, evolution has already passed more than half of the way from Nothing to man. For example, a RAM is already almost 90% human.

10.6689.

The fact that the mushroom's DNA is 55% human-like tells us that the hardest part is getting started. The road from a mushroom to a person is an easy 45% of the way.

10.6690.

The weak evolve through symbiosis, which is metaphorically love and synergy. The strong evolve at the expense of parasites and viruses. Strong in the process of fighting viruses, fungi and parasites integrates their DNA into its DNA. In human DNA, for example, about 30% of the DNA of viruses.

10.6691.

The strong weakly strives for symbiosis, because symbiosis is a Union of the weak in order to survive and become stronger. Strong so strong, they have no special reason to unite.

10.6692.

Blessed are the poor in spirit, because they are capable of love, unions, symbiosis, and synergy. Synergistic effects increase the combined strength of the weak by thousands of times.

10.6693.

The weak are more capable of evolutionary growth than the strong. The weak are able to unite, consolidating their best evolutionary advantages and findings.

10.6701. Meat apples.

Worms and wood are an example of a tricky symbiotic Union. The tree is very profitable to produce meat apples.

10.6705.

The earliest inhabitants of this planet are anaerobic bacteria. Second came their deadly enemies, the mushrooms. The third were viruses. However, it is possible that the viruses were the first, because they are not living viruses. A virus is a transitional form of life from the inanimate to the living. Human DNA is 55% similar to fungi, 30% to viruses, and the rest is probably due to bacteria.

10.6706.

In fact, the evolution of life took place in three stages. First there were viruses, then anaerobic bacteria, then fungi. The mushroom is very similar to the crown of evolution, because all the other stages of life development are already a superstructure over this Trinity.

10.6708.

The vices of the viruses. People are living cells. A virus-infected cell signals that it is infected. Such a cell would have to "commit suicide" in a process called apoptosis or programmed cell death. One of the main functions of viruses in nature is to control animal populations. The higher the population density, the more actively viruses attack it to destroy excess individuals, reducing intraspecific competition and the burden on the food ecological niche.

10.6717. The Holy Trinity.

The battleground of planet Earth is a war between fungi, bacteria and viruses.

10.6718. The Cambrian explosion.

The invention of the chord and skeleton removes the size and strength restrictions on multicellular organisms. Firmness of character is very useful for the body.

10.6723.

God's gift is love. What a man falls in love with, he will turn into. Therefore, smart people are very attentive to the objects of their love.

10.6724.

What is love? This is the mode of information exchange. You took something you were really excited and it's handed to you on a new DNA information and changed you... Now you intuitively want to be like the object of your love... you study it, take the best out of it, and build it into your mind genome.

10.6730.

The narcissist is an ever-diminishing entity that strives to the limit of zero. Turning into a dot is very painful, so daffodils are very nervous and suffer greatly.

10.6731. Pulsar.

The essence of the pulsar is that at first a person falls in love with something external and grows. Then the idealist becomes disillusioned with the object of his love, and begins to love himself. The love of self causes him to shrink back to the point. Grow-joyfully, shrink-sadly. In other words, as long as you love art, you grow, but as soon as you start loving yourself in art, you instantly start falling into hell.

10.6739. Are there only idiots around?

If you love the high, you will start growing. If you love the low, you will begin to shrink. They say you need to love people. On the one hand, this is a good idea, but on the other, it will paralyze your ability to grow above the average level of your surroundings.

10.6742. Who are you?

When you firmly love one thing, you turn into that one thing quite quickly and predictably. When your love is smeared on a lot of things, you take a little something from everywhere, as if collecting yourself, like a mosaic, one piece at a time. The first option is well predictable, and the second option will theoretically allow you to become yourself.

10.6743.

A person is inclined to love what surrounds him and what he has. Tell me who your friends are, where you live, what you have, and I'll tell you who you are. A person automatically turns into something that they love, hate, or just pay attention to.

10.6752. Free people.

In a three-class society, there are lower bacteria and single-celled ones, which are dominated by a complex hierarchy of fungi, headed by a higher mushroom. The third force is viruses, those who did not want to live in the system of slaves and masters, but they can not live without them. Viruses are nihilists and free people who don't like this whole system of slavery and want to destroy everything.

10.6754.

It's good to be big. The big one can eat everyone, but no one can eat him.

10.6755.

Life is a symbiosis. There is no life without symbiosis. To survive, we must form a single symbiotic organism. The life of a single-celled creature is unbearable. Anyone can eat it. We must create a multicellular symbiont society.

10.6763. Don't deviate from your goal.

Observing the evolutionary processes, I noticed that nature never gives up. The idea to create mushrooms was attempted several times, and ended in extinction. Only 3-4 attempts were successful.

10.6768.

The causes are not important, the consequences are not important. Nature and evolution are interested in processes occurring here and now. It is important to survive, reproduce and create now. What happened yesterday and what will be there tomorrow-we are not interested in much. The past has already taken care of itself, and the future is well spoken of in the gospel.

"So do not worry about tomorrow, for tomorrow will take care of its own: enough for each day of its own care."

10.6771. One egg for all.

The devil is a kind of virus. Viruses are needed to thin out the population density... Something like the attendants of the forest... But then again, when trees grow too close, it is also stupid and ignorant. I had to crawl away, find my place in the sun. And these are huddled together in fear, all identical, like identical twins. Now we have to thin them out.

10.6775.

Big melts the one who is not afraid. Fear limits things in size. The less fear, the more growth.

10.6780.

It is not the individual that evolves, but its DNA. DNA is the host of the body. DNA evolves and changes the body for itself. First you change yourself from the inside, and then everything changes from the outside.

10.6796.

This world is perfect and only the perfect has value in it. The main way to create perfection is uniqueness and novelty. Because you are the first, you are automatically the best.

10.6802.

Death, fire, viruses and parasites are needed to turn a lot of small coal into a little large one. The evolutionary enlargement of forms requires free space. A lot of people – not enough oxygen.

10.6811.

The point of dislike is to preserve your uniqueness. We become what we love, and this destroys our uniqueness and perfection. However, it is impossible and problematic not to love anything. It is optimal to love everything at once, this will make your own uniqueness more harmonious.

10.6812.

Love for idols allows us to grow well and quickly, borrowing the best from them. When we grow up, this love will break, and we can fall in love with something else, gradually accumulating our uniqueness brick by brick.

10.6815.

You're not immune to what you don't have. Poison in small quantities creates the immune system. Without immunity to poisons, you can't live in this world.

10.6822. Fly in the ointment.

A pinch of poison is immune to poison. Beauty is health. To be healthy, you need to be a little sick.

10.6828. New truth.

To create a new self, you need to completely destroy the old self and completely update your information system.

10.6829.

The tree is reborn in its seed. Having got rid of material and superfluous information, the tree puts its soul in the grain and finds a new life.

10.6833.

Learning is a process of destruction. The process of growth and creation is the experience and application of new knowledge in practice.

10.6837.

The one who is not afraid grows big. Fear limits things in size. The less fear, the more growth.

10.6851. A sacred ritual.

Life is movement. Experience is movement. If we build a movement, we will create life. It is said that every day has its own concerns, it is about movement. We must move at all costs, nothing else matters.

10.6854.

You turn into something you love. If you love a zombie box, you turn into a zombie. If you like cartoons, you turn into a cartoon. You love dogs, you turn into a dog... Should I go on?

10.6858.

The state of dependence preserves the personality. To acquire the ability to grow, you need to get rid of all the helpers and saviors. It is terrible, of course, but this fear is an illusion.

10.6888.

Pride is contempt for the inferiors. The devil wants to be the best and despises everything small. Real angels are tiny accidents, specks of dust and atoms. God is at the bottom, not at the top. The universe grows from a point, not degrades from somewhere above.

10.6890.

Enlightenment is a state of absolute clarity, when the grain is completing the programming of their DNA under your goal. Now that everything is absolutely clear, you can start growing.

10.6891.

The philosophy of Syntalism predicts all of humanity in the coming years a hundred ... complete extinction ... through the dissolution of virtual computer illusions in the worlds. Just a couple of decades of the development of technology and all to hell will perish in paradise ... Some righteous syntalists will remain, for someone must inherit hell ... Right? ... I like it in hell ... we have good hell. .. pleasant in every way ...

10.6895. Experience of knowledge.

Love is the knowledge of one's goal, and experience is growth and movement toward it.

10.6901.

The strength of Syntalist is that it has love and admiration. Love protects him from temptations and vices. Love allows him to admire and love his enemies, competitors and, thereby, to borrow all their strengths and unique sides.

10.6902.

Love your enemy, for then you will be stronger. If you love your enemy, you will learn from him. You will take from it its most powerful and useful, learn the secret of its power and power. Knowing your enemy and becoming stronger, you will solve all your problems.

10.6903.

We are persistent people, we are not sinners. We know how to never give up under any circumstances. 1% percent of effort per day during the year gives 300% growth. And if these efforts, for example, 12%... and let's say for 50 years... Will you count it yourself?

10.6904.

Truth, faith, hope, and love... this is a monstrously terrible force. As long as sad sinners dream of Paradise and give themselves up to vices, we will be able to realize any of our dreams...

10.6905. Knowledge is power.

The right syntalist borrows all the best and strongest from his competitors, for he admires them. There is no pride in the syntalist ... He does not judge, he loves the truth, not himself in the truth. The philosophy of Syntalism today has the most holistic and complete understanding of the structure of being...

10.6906.

Love is when you take information and grow. Pride is when the opposite is true. Pride is a mode when information is denied and growth becomes impossible.

10.6917.

According to Syntalism, man was banished from Paradise for humane reasons, because Paradise and reason are incompatible. In Paradise, the mind degrades and dies. The mind is that which overcomes problems and obstacles. In conditions of complete relaxation, higher fungi degrade and turn into single-celled fungi, a typical example of this is yeast.

10.6920.

Fear is what hinders growth. Fear is a lie and an illusion. If you get rid of lies and delusions, the fear will disappear, which will provoke explosive growth.

10.6925. Vice is the form, virtue is the content.

The child loves his parents, and if he loves them, he learns from them and receives information from them, on the basis of which he builds the DNA of his mind. If there are many vices in the parents, then, first of all, the child learns them.

10.6926.

When you love and admire a person, you learn good things from them. When you are afraid or envious of him, you learn bad things from him. From the first you grow and rejoice, from the second you degrade and suffer.

10.6928.

Blessed are the poor in spirit, because they do not consider themselves the smartest and are ready always, everywhere and from
everyone to learn, adopting their best and unique.

10.6940. Unity and struggle of opposites.

Love your enemies. If your fight is long enough, you will become
them, and they will become you. War and conflict will bring you
to unity.

10.6961.

In conflicts and disputes, we train our skills and become stronger.
Barriers make us stronger.

10.6963. Christmas tree.

Syntalism is a philosophy of creation through destruction. The
meaning of Syntalism is not to let the world stop. Life is a cycle of
movement of destruction and creation. Idols want to stop moving and don't want to die. The Syntalism philosophy gives the
new generation the power to destroy the old generation, thus sav-
ing the world from being stopped and degraded.

10.6969.

To create the universe, it must be destroyed. To create a new self,
the old self must be destroyed.

10.6972. A hero without a feat is sad.

The hero, anticipating the fight, is happy. The hero loves his enemies, and resistance and obstacles only make the hero stronger.
The more resistance the hero meets, the stronger he becomes and
more happy.

10.6974.

Tenderness leads to fear, and fear leads to vices. By eliminating
tenderness, you become immune to Vice.

10.6979.

When you meet evil, don't judge it, but love it. Judgment breeds fear and lies. Love breeds knowledge. Having known your enemy, you will learn how to protect yourself from him, and you will also be able to take the best of his properties for yourself.

10.6980. Love is Aikido.

Human DNA is 30% made up of the DNA of the viruses that attacked it. Getting to know your enemies, a person uses their weapons against them.

10.6981. Tell me who your enemy is and I will tell you who you are.

To grow and become stronger, you need enemies. Knowing your enemies, you take their weapons. Your enemies are the source of your power and energy. A person's DNA is 30% made up of the DNA of viruses that wanted to harm them.

10.6983. John barleycorn.

The strategy is as follows: choose a more complex and more beautiful goal. Your goal is an enemy to love. Love is knowledge. Knowing your enemy, you will learn a lot from him. The more complex your goal, the more it will teach you, and the stronger you will become. When you know your goal, this information will change the DNA of your brain. The altered DNA will change you, too, and you will be reborn as someone who can achieve any difficult goal. The main thing – do not be afraid, do not be nervous, do not rush, hope for success. Love is God and truth. If you have love in you, gradually the truth will appear, it will turn into faith, and faith is the greatest power on our planet.

10.6984.

To be better than your enemy and defeat him, you must admire him, not despise him. When you admire, you know it, and you are able to defeat the enemy with its own weapons. When you despise your opponent, it creates fear in you and makes you weaker.

10.6987. Tell me what you want, and I'll tell you who you are.

You need a challenge. To realize your goal, you must know it. The more complex and beautiful your goal, the more perfect the DNA of information you will acquire by working with it.

10.6988.

You can't grow without a goal. When a person learns his goal, he reprograms the DNA of his brain and, guided by this new information, begins to grow. There is no goal, no new information, no growth.

10.6994. Self-loathing.

The point's desire to become the universe is not pride, but a normal evolutionary process. Pride is the contempt of the universe for the point from which it arose.

10.6995. Look for benefits.

The meaning of problems and mistakes is to guide a person on the right path and make them benefit from the situation for their life mission.

10.6998. The yeast fungus.

Murderers evolve, the evolution of victims is secondary and based on the knowledge of their enemies. If you destroy the killers, enemies, and problems, evolution will end, and the victims will degrade into single-celled yeast.

10.6999.

Evolution is not a constant, but a dynamic process. Species either progress under the influence of problems, or degrade under conditions where everything is good and fine.

10.7000.

Paradise consists of the simplest single-celled creatures, so when you have no problems, you start to endlessly degrade.

10.7001.

The devil is all so complex and perfect from the fact that he lives in hell, where someone constantly wants to eat him.

10.7002.

Since the main engine of evolution is inanimate nature (viruses and circumstances), we can say that the cause of evolution is accidents, errors and space-time limitations of existence.

10.7004.

The meaning of the needle's eye metaphor is that truth and angels are the simplest and most banal. Paradise is a world of single-celled immortal beings. The more perfect and complex everything is, the more hell there is.

10.7005.

Civilizations are dying out for two reasons. Or they will degrade into a single-celled Paradise. Either technology and complexity reach the point where hell becomes unbearable. The unbearability of life in hell leads to the self-destruction of the system.

10.7023. Three assholes.

There is no heaven in the real world, but there are three kinds of assholes. Ass-clam is when a ringworm evolves into a shell and lives inside its shell, escaping from reality. Clams are primitive, weak and tender.

The second type is the ass of boiled crayfish, which are cooked in hell and suffer very much for this reason. This is how idols and trees live.

The third type is an ass filled with humility, which is satisfied with everything and knows what to do and why. This is how all the higher animals and people live.

10.7025. Evolution of the sperm.

There are three ways for worms to evolve. Ringworms (a squirrel in a wheel) evolve into mollusks and snails. The other two paths are the evolution of earthworms into humans and other animals. You can evolve without endlessly suffering, this is the second way. And you can put up with pain and suffering, turning them into joy and excitement, this is the third way. The third way turns the sperm into a human, and the second-into a tree or animal.

10.7029. Single-celled joys.

Demons despise angels because angels are primitive, like infusoria shoes. Paradise life turns angels into cyanobacteria that feed on the sun. Demons have something to be proud of, demons are mushrooms and idols, they rule over bacteria, turning them into their slaves, they feed at their expense. But bacteria do not care, they live in an eternal Nirvana disconnected from reality.

10.7030. Look for problems, problems are the fuel of growth.

Avoid Paradise. Paradise is a place of degradation, where everything is destroyed. Destruction and degradation are so painful that even rotting alcoholic pleasure does not save the situation.

10.7031.

The victims are encouraged by viruses to improve their weapons. The more appetising the victim, the more you want to eat it.

10.7032. God's sheep.

The symbiosis of sheep and wolves is that the stronger the wolves become, the more appetising and numerous the sheep become. Viruses train and contribute to the evolution of their victims. The more effectively wolves attack sheep, the more effectively sheep evolve.

10.7035.

Life in Paradise inevitably dies, because shivers, eating sugar, produce alcohol, which then kills them.

10.7044.

Since the attainment of truth occurs through the denial of false-hood, therefore, in order to attain the truth, conflict is needed. Creating a conflict. We fulfill the commandment "Love your enemy" and thus learn the truth.

10.7072. Experience of knowing the truth.

When you are not afraid and do, you start learning ten times faster.

10.7073.

Curiously, viruses and diseases are more to blame for human evolution. Having mastered the fire, the person became quite tender and painful. More diseases-stronger evolutionary progress.

10.7074.

Black from white is about as different as a monkey from a human. That is, it seems different, but the evolution was so gradual and imperceptible that you can not say that once a mother APE gave birth to a son of a man.

10.7101.

Running away from problems and fears is dangerous. You can turn into a clam or yeast. Clams are very stupid, weak and cowardly.

10.7104. Training mode.

In the absence of problems, enemies and fears, the brain begins to invent them. Otherwise, it will degrade and die.

10.7127.

Truth is born in dispute and conflict. This does not mean that the parties will come to peace, convince each other, or one of the parties will definitely win. This means that one or more truths will be born.

10.7149. Immune protection.

It is best to protect yourself from the enemy with their own weapons. Than it attacks you, pull it out, and shout: "This is mine!"Viruses are powerless against their weapons.

10.7210.

The point of an evolutionary niche is to learn to eat what others don't eat.

10.7567. RNA heredity.

The effect of genetic inheritance is minimal. A human is a stem cell whose purpose depends very much on the circumstances and instructions of the RNA.

10.7671.

Practice is a form of theory evolution. Practice is the acceptance of the theory with reality, the process of removing everything superfluous from the theory.

10.7834.

Boredom is a great thing, if it weren't for boredom, man would still be a monkey. The main thing is not to interrupt boredom with entertainment. Work and useful things, that's the best cure for boredom.

10.7968.

What is revolution?

- Destruction in the name of construction

10.8220.

Love gives rise to three possible scenarios. Desire to turn the object of your love into yourself. The desire to become the same as what you love. The desire for symbiotic unity. The third is co-operation, and the first two generate the threat of fear, although if not fanaticism, they are also useful, because they allow you to exchange information and evolve.

10.8251.

The meaning of the phrase "simple as three rubles" means that energy is the simplest and most primitive form of energy. The complex tends to disintegrate into the simple, and the simple tends to evolve into the complex.

10.8643.

In monkey communities, male power is based on the support of females. Females can't stand each other. From this point of view, it makes sense for young males to attack older females in order to earn the love of the young. On the other hand, old females have a lot of power, but they are very lonely and no one likes them. The latter fact opens up particularly great opportunities.

10.8711. They say that God created man.

Initially, all people were black, but in a cold climate, black people can be cold ...and they turned white... and in the wet they turned yellow, for there were also very good reasons for this. Given that God is a reality. It turns out that they were separated by God, forcing them to adapt to the external conditions of life?

10.8814.

In monkey communities, male power is based on the support of females. Females can't stand each other. From this point of view, it makes sense for young males to attack older females in order to earn the love of the young. On the other hand, old females have a lot of power, but they are very lonely and no one likes them. The latter fact opens up particularly great opportunities.

10.9003.

The new always comes from the negation of the old. The new and the young come, make a revolution, and burn all the old. New truths, in order to clear a place for themselves in the new world, are forced to burn out their living space with fire and sword. The old one clings to life and doesn't want to just die and give up its energy. The old is mired in pride. Moreover, it prepares its successors by force and tries to impose its ideals on the future by force. The devil buys souls and promises Paradise. The old one passes on power and energy to his minions in order to preserve his agony and lies in time.

10.9326. God is fire.

Fire is a method of development, evolution and formation of human civilization. We are not talking about a metaphorical God, but about the essence of physical fire and its influence on the development of historical processes. If the question is interesting, you can find my book "History of religion" on the Internet... it describes in detail the mechanics of the influence of the fire cult on the development of society, technology and science.

10.9440. Useful evil.

Beauty is a drug, you get used to it all the time, so you need more and more new forms of beauty. The beauty of this virus, it is a predatory parasite that turns the brain into a slave using love as a chain. Protecting itself from beauty, the brain develops an immune system, making beauty part of its DNA. Beauty is like a virus, constantly mutating to successfully attack its victims. Beauty victims evolve with their killer, turning themselves into killers. He who conquers beauty becomes beauty himself. The one who loses to beauty will die.

8.1902.1.

An interesting thought about cheating.

Of course, female cheating is bad and full of big problems for women themselves. But let's see why it happens. And why nature gave women such an opportunity.

A woman can cheat not for nothing but for some ego boost. It's not just anyone she may choose but only the best of the available "stud muffins". Of course, a "wild stud" will never marry her and will never become a domestic draught horse but she likes his fresh blood and good genes. A woman can easily fall in love with such a stud and may get pregnant from him. Nature craves for genetic diversity and the culmination of best people. And the fact that a woman will later give birth to a cuckoo fledgling and her husband will have to feed them all, coincides with the idea of working bees and ants. It's genetically best beings who provide future generations while others are meant to work and feed them. Such state of things coincides with the evolution ideas about maximal distribution of the most perfect or diverse genetic code. As for the diverse genetic issues, it can be noticed that the more genetic differences there are between a man and a woman, the better it is. In ancient times it looked the following way. One day some traveler came from afar and stayed on the occupied territory without any means of survival. But his genetics seemed useful for the local individuals. Some female liked this traveler and got ready to get pregnant from him so that the old male in the local community would feed the newborn and let it raise on his territory. Thus, he would become the owner of resources who would feed "novelty" and "fresh blood". Or "perfection", as women like to deal with the best of available males like bosses or some "people person" or unofficial leaders.

Thus, seen from morality and family perspective, female cheating is a bad thing but considered from evolutionary ad genetic perspective it's a good thing.

10.10147.

Guilt should be accepted and enjoyed in the sense that it is the cause and support of our evolution and personal growth. In the absence of guilt, there is no reason to change.

10.10245.

The black square is not the beginning of art, but its end. The black square is the ultimate chaos in which there is no more free space. A black square is a billion superimposed shapes. The black square is a beauty that has reached its limit.

10.10281.

The reason for the appearance of man is very accurately described in the Bible. Once the worms and the Apple tree entered into a symbiotic Union to make apples more valuable to animals. Wormy apples added protein to the diet of humanoid monkeys, which gave an excess of energy useful for brain development.

10.10683. Unisex.

For corporations, a woman is more profitable than a man. Since the world is ruled by corporations, which are the main source of energy for people, this generates evolutionary processes aimed at increasing femininity in males.

10.10780. Self-generating system.

A reasonable way of thinking forms the paradigm through which a person observes the world. Consciousness defines a paradigm, and being back through that paradigm defines consciousness.

10.10834.

A person is able to love, that is, create new habits and conditioned reflexes. Animals can not love, all their habits and behavior are due to innate instincts and reflexes. A person can change himself from within by the power of the spirit, animals cannot reprogram themselves.

10.10917.

Many people are not able to see beauty and God in all its diversity, which removes them from God, but brings them closer to animals.

10.10948.

Although the complex consists of the simple, it is not the sum of them. Each new evolutionary state of a complex is independent and distinct from its old States.

10.11067.

The peculiarity of man is that he is not the crown of evolution, but his mistake, that is, a cancer cell. And, of course, the immunity of being tries to surprise a person, poisoning his life. The main idea of all religions is that if you go back to God and start doing everything right, then there will be bliss and happiness. Let's just say that happiness may come, but the mind will die, and the person will turn into a blissful idiot. Syntalism offers not to kill the mind, but, having accepted pain and suffering, to develop immunity.

10.11189.

Syntalism believes that many of the laws of nature are directed against man, because man by nature is extremely similar to a virus that antivirus software will sooner or later destroy. On the other hand, we are very happy that there are clever ways to circumvent the law, not fall under it, and again... this law will work if you cross the line, but sometimes you can lie low and wait. Viruses have been fighting fungi and bacteria for billions of years, and they are still the most alive... On the third hand, although a person looks like a virus, it would be more correct to define it as a Trinity of viruses, bacteria and fungi.

10.11222.

When something grows, it is an evolutionary development. Revolution is when everything stopped, froze, and then started again from scratch.

10.11224.

Animals, having found their own way to get energy, turn their meaning of life into food. From food, animals get fat and become the prey of other predators. In this system, a person who strives for perfection becomes the most powerful predator, devouring everyone. On the other hand, the perfect man is merciful to his victims, for this is his strategic resource.

10.11433.

Everything is perfection, but every new form of perfection is not the evolution of the old, but the creation of a new form. Both forms continue to exist simultaneously and fight for energy.

10.11692. The elephant and the gnat.

The more imperfect a thing is, the more likely it is to survive in the short term. The more perfect a thing is, the greater and longer its victories.

10.11730. Perfection is a dynamic variable.

Who is perfect? It is the one who survives and wins the natural evolutionary struggle. Wins and multiplies. Perfect is the most adapted to life, that is, to the current conditions. There is no limit to perfection because external conditions are constantly changing and perfection changes in sync with it. Perfect requires great care and love for reality in order to have time to react quickly to all its changes.

10.11731.

The perfect one is one who honors the law of natural selection. The perfect one is the one who wins the evolutionary struggle, adapting as much as possible to external circumstances.

10.11732.

The law of natural selection significantly restricts the freedom of human will. You can want anything, but if your desires weaken you, you will be eaten or you will not be able to get food and reproduce.

10.11733.

The laws of natural selection have less effect on a living cell than on a large living organism of which this cell is a part. The whole protects the small details that are part of it from natural selection.

10.11742.

Animals are parts of the ecosystem, and their evolution leads to a deepening of their specialization. A person is a separate and integral system that is capable not only of specialization, but also of harmonious system-wide development.

10.11744.

Parts of the whole tend to lose their individuality. Individuality is a property of complete and independent systems.

10.11828.

Pride is a universal force that pulls us to perfection. It makes the strong beat their heads against the wall and seek victory in the rat race. The weak are looking for uniqueness and competition-free territory. By eliminating pride, you will stop progress and evolution.

10.11872.

Vices make you inattentive and hinder your perfection. Perfection is a victory in the evolutionary struggle, the maximum adaptation to reality, which is impossible when you are inattentive.

10.12291.

It can be assumed that the main reason for the evolutionary backwardness of Papuans from New Guinea is Batel Nath. The same effect on the people of America to produce Coca leaves, and on the Eastern residents – hashish and nasvay. You can also remember alcohol, drugs, and all other ways to escape from reality.

10.12391.

Those who are courageous are not afraid of their surroundings. The point, crushed by the ocean of darkness, explodes and turns into the universe.

10.12418.

Studying biology, I learned that a fish is almost like a person.

10.12419.

The most primitive are insects. Reptiles eat them, further evolution of reptiles turns them into birds. Metaphorically, reptiles are demons, and birds are angels.

10.12420.

Which came first – the chicken or the egg? There used to be caviar, fry, then reptiles, and then chicken.

10.12724.

In animals, attention is controlled by the carrot and stick, but intelligent people can arbitrarily control their attention and effort.

10.12799. DNA of Ideas.

An idea is a virus, a disease that infects reality, changes it, or even tries to kill it. I am glad that having had an idea, reality develops immunity to It. It's frustrating that viruses are constantly evolving. Interestingly, virus DNA makes up up to 30% of the DNA of reality and significantly changes its existence.

10.12984. Ocean of desires.

If we metaphorically compare desires with the ocean, then the evolution of man is the appearance of fish on land, becoming Buddhists and liberation from the bondage of desires.

10.13040.

A real person is a virus whose purpose is to look for vulnerabilities in the ideal algorithm of being, thus improving being. God is an ideal order that man always struggles with and periodically wins.

10.13041. Evolutionary determinism.

It is a delusion that the perfect and the strongest win the evolutionary struggle. In the evolutionary struggle, the weakest win, and the strongest die out in the struggle with each other and with themselves.

10.13042. The winner of the weakest.

You strive for power, believing that it will help you win... you don't know yet that the weakest wins and survives. Blessed are the poor in spirit.

10.13058.

The only thing that can save a proud person from hellish suffering is work and striving for improvement. Fixated on work and personal growth, the proud man will save himself from the temptations of vices and mortal sins.

10.13066.

You don't have to go up all the time, you can go straight. You can even go left, right, or back... It all depends on what, where, and why you need it.

10.13098.

We can only learn what we don't know ... what we know can no longer be learned.

10.13099.

It is useful to read the biography of a worthy person and, associating yourself with him, take his virtues for yourself. Of course, vices are easier to associate, but you choose the hard way.

10.13226.

Imperfection is a Vice. It takes time to achieve perfection. Don't get too busy. Don't take the new one until you're perfect in the old one.

10.13264.

Evolution of the addict: sweets, fairy tales, games, movies and TV series, alcoholism, drug addiction, religion. If the addict does not disappear in the intermediate stage, sooner or later he comes to seek solace in the Church.

10.13440.

Having moved to the savanna, the monkey, in order to fend off predators and hunt, was forced to develop a mind that could invent weapons. There's nowhere to run from predators in the savanna. Running prey is also not catch up. We had to invent sticks, spears, traps, bows, etc.

10.14174.

From the point of view of psychology, the soul is a software package, and the brain is an electromagnetic computer system. So you can think of your soul as a virus that has enslaved the poor monkey's brain.

10.14451. Civilizational growth.

Overdose of sugar and pleasure, reducing the overall sensitivity of the body, while making a person vulnerable to acute sources of emotions that carry novelty. Thus, the chronic depression of citizens supports the demand and economy in the state. Economics and trade stimulate science and culture. Thus, evil begets good.

10.15424.

It is a lie that primates (monkeys) were vegetarians, for they ate worms, insects, and eggs wonderfully. The idea that people can do without animal squirrel has no historical roots.

10.16529.

What you call a loss of control and a breakdown is switching to autopilot. Switching to autopilot occurs when there is no clear conscious understanding of what to do. Autopilot includes instinctive programs of the subconscious, tested by millions of years of evolution and have shown their effectiveness.

10.16821.

God is love... having learned to love, the monkey found reason... Animals don't know how to love... However, most people do not know how to love, which, of course, excuses animals and makes them look like many people.

10.17057. A perfect mind.

They say that the human mind is the crown of evolution, the strongest of the strongest. This is a delusion, the crown of evolution is God.

10.17109.

There is no better or worse, only evolution. There is growth, and there is withering.

10.17373.

There are many gods, but each has its own God who leads it. God is an idea, everyone has their own ideas. Man follows his ideas through life. God is a forest, but every monkey lives on its own tree. The forest consists of trees.

10.17624.

Light is the initial form of hardness. When light grows, it becomes solid and material. The entire material world is the evolution of light.

10.18278. Beats, means, likes.

The point is that if you can't love and make friends, then you need to escalate the conflict. He who does not desire unity in love is the enemy. The enemy must be loved. The stronger the enemy, the better it is. As we learn about the enemy, we adopt their best DNA and evolve. To sum up: if love does not stick, provoke scandals. Scandals is a good thing. For the wave nature of neurotic relationships, scandals are very life-giving. Movement is life.

10.18482.

Difficulties are like viruses, overcoming their object, absorbs part of the code of their DNA and evolves. The stronger and more perfect the difficulty, the more useful it is.

10.18488. Ideas produce ideas.

Which came first – the chicken or the egg?
One idea gave birth to another, the second grew and gave birth to a third. Over time, ideas hang down and evolve. The very first idea was very simple, the very last idea will also be the beginning.

10.18511. Symbol Of The Devil.

The symbol of the serpent (dragon) devouring its tail can be associated with the symbol of the squirrel in the wheel and imagine the meaning of life as an endless improvement and renewal of being. The more perfect eats the less perfect, and so the system evolves.

10.18610. Just hope for yourself.

You've been relying on others all your life, and it's filled you with fear. To overcome fear, you need to repent, that is, change your mind. When fear disappears and becomes love, the world of illusions collapses. And that's what they call Armageddon. Love will give birth to a new sprout of real life, small at first, but eventually it will grow stronger.

10.18962.

Once upon a time there was a monkey. Until her brain was infected with a virus called "idea." Ideas are viruses responsible for the emergence of the human mind. Through his ideas, man looks at the world, and being determines his consciousness and life.

10.19221.

The main motive of the story is human pride, that is, thirst pleasure, sublimated into a lust for power, money, and sex.

10.19329. Evolution of thought.

We can say that the book Variothoughts is written in the genre of real RPG. If you read the book sequentially first (Volume 1, 8, 9, 7,6,5, 4, 3, 10), then you can watch the evolution and growth of the perfection of thought together with the author. The reader gets a unique opportunity to observe the evolution of thought from heroism through nihilism and mysticism to realism and pragmatism, and by joining this growth, evolve with the book.

10.19405.

Property, real estate, and money are the material embodiment of time. Time is the simplest form of energy that continues to evolve and transform into matter.

10.19477. The immutable law.

The evolution of similar systems is the same. The laws of evolution are a constant. A woman always chooses the same type of man, because she does not change. She constantly chooses A, which grows and evolves into B, but she does not like B and again chooses A. the Circle closes. However, men who are hungry for love, think similarly.

10.19530. Absolutely black.

Black is the evolution of white. The perfection of the whole, breaking up into innumerable forms, becomes absolutely black. White is just one form, and black is our whole world of forms and illusions. Black therefore has zero energy charge, which gave all the energy to being. The absolute black.

10.19647.

Truth (absolute) this is a kind of integrity that has reached a critical mass and divided itself into an infinite number of new forms. It is impossible to know the truth, because the truth is the whole real world, which is infinite, because the division of the absolute can take place indefinitely. In fact, what you call the evolution of the system is the continuing degradation and disintegration of the system from a simple and unified absolute to some sophisticated and very complex forms.

10.19659. Divide and conquer.

What you call evolution is the disintegration and division of the whole into many things. The whole, expanding outward, absorbs simple energy and orders it within itself. What you call order is division. Order, dividing the streams of energy, dominates them.

10.19663.

Death is not the end, but renewal. The real world is something like an immortal multicellular being, where there is a constant process of updating living cells.

10.19826.

Patience is a matter of evolution. Revolution is a product of pride and impatience.

10.19827.

Evolution inevitably results in revolution. Evolution is the process of gradually reaching a critical mass, the initial stage of an exponential curve. Upon reaching a critical mass occurs and the detonation of the explosive growth of the system – it's called a revolution. Evolution and revolution are in dialectical inseparable unity.

10.19849.

The critical mass of a system is a point after which even a small increase in the system outside begins to generate a significant division within itself and expansion in other dimensions. The system gradually evolves from one-dimensional to three-dimensional and generates an explosive energy flow that could be called time or money.

10.20003.

The meaning of free will is that a person can determine their own purpose in life. Having defined their goals and fallen in love with them, a person begins to learn them and learn, gradually rebuilding their internal structure for these goals. As you achieve your identity and your purpose, your life will begin to change in the right direction.

10.20225.

Man is an insectivorous APE, and in the worm Apple he is more interested in the worm than the Apple. An Apple is either an appetizer to a worm, or something unnecessary, harmful and useless.

10.20238.

Love your enemy as the RAM loved the shepherd and became one of the dominant forms of life on the planet Earth. Chickens, pigs, and dogs also love their owners.

10.20357.

The system is designed so that the object produces food, which is constantly claimed by parasites and consumers. Parasites benefit from the long-term existence of the energy source, so they take care of its protection and reproduction.

10.20358. An unrestrained parasite.

Parasites protect the system in such a way that they do not allow other parasites to enter it, which, being uninterested in long-term cooperation, would like to eat everything at once and so destroy the system.

10.20359.

A humble parasite drinks a little of its own blood and drinks, such a parasite is good for the system. The greedy parasite attacks, tries to destroy the system and quickly devour everything. From such parasites, the system is protected, among other things, with the help of the first restrained parasites.

10.20363.

Any parasite that takes care of protecting and reproducing its source of energy gets this source of energy at its disposal. You need to love your energy sources.

10.20367.

There is no struggle for a place in the sun. Point already exists. In the future, accumulating information and reaching a critical mass, the point begins to decay inward, so turning into the universe.

10.20735.

That which has lost fear is formless. Too much fear creates real estate and death. Moderate fear is movement and diversity of life forms.

10.20740.

The real idea is impracticable and like hope. Hope is one of the names of truth, and truth is unknowable. The idea evolves endlessly. You follow your idea, and it follows the infinite path of perfection.

10.20867. Goals give birth to means.

We cannot say that man is a Creator, but man is a creation... A reasonable man is a creature of gratitude, grace, and nobility. Man is a tool that creates. Just as man creates an instrument to achieve his goals, so God created man to achieve his goals. A person is a means that creates other means by which certain ideas are realized. I believe that an idea, wishing to realize itself, creates a whole chain of production facilities, one of which is a person.

10.20883.

Man is not the crown of evolution as such, but a fairly perfect tool with which nature has discovered a new way of creating new and unique forms. Those forms of the real world that are generated by man are a new evolutionary step in the development of the universe.

10.20901.

Shepherds, dogs, wolves and sheep are a symbiotic system. The shepherds need wolves so that the sheep are afraid of the wolves and seek salvation from the shepherds. Dogs this is a situation when a shepherd, having loved his enemy, turned a couple of wolves into his friends, and now they protect him from other wolves, and help manage the sheep. In principle, the rams are also all profitable, having reconciled with their enemy, they, taking advantage of his predatory greed and insatiability, have become one of the successful biological species. The shepherd is a dominant predator, which, by protecting its food from other predators, will allow the sheep to occupy a solid ecological niche.

10.20903. Spoiled food.

Parasites are very useful for their victims, because they protect them from other predators and parasites. Symbiotic parasites kill their victims for a long time, allowing them to reproduce and live. However, getting into foreign organisms that do not have protection from this parasite, they quickly destroy it. Thus, extraneous predators avoid eating other people's food, feeling a threat to their own lives.

10.20904.

The idea (love) this is a predatory parasite that turns a person into its slave, while protecting it from other even more dangerous enemies. Love is a predatory and cruel parasite that requires unquestioning loyalty, but it gives joy and a lot of other useful things for the body. Love protects its slaves. All your other enemies will just eat you up and make your life a living hell.

10.20931.

To find food, you need to look not for food, but for someone who will eat you. A predator that devours you, it also creates your ecological niche and protects you from other enemies, including yourself.

10.20936.

If you want to eat someone, offer them protection. Predators must take care of reproduction and protection of their victims.

On the other hand, if you need an evolutionary niche, find some-one who wants to eat you and is willing to protect you for it.

10.21102.

The Trinity of the father, son, and Holy spirit is not separable. The father is energy, the son is the vehicle of energy, and the Holy spirit is the forms of the real world. Man is an incarnation My son. One might say that it would be possible to do without man, but man is not an independent and finite entity. Man is an evolution-ary form of the essence of a semiconductor (transistor, micropro-cessor, faith, love, order, God). Man is one of the representations of the entity that controls the flow of energy between Father (Nothing and an ocean of energy) and the Holy spirit (real world).

10.21103.

Man's self-respect should be based on the fact that he is an evo-lutionary form God. Man is the embodiment of the transistor (microprocessor), law, and love. If God is a huge cloud computing system, then man is a transistor, a coprocessor, or a local com-puting station. Man is equal in value to the ocean of energy (God the Father, nothing, non-existence)... and equal To the Holy Spirit (being and the real world).

10.21143. It's very simple.

Just do what you're doing and throw it in the water. It's not your business to think who will pick it up from the water and why. You feed your energy from the earth and the sun. Those who eat your fruit are parasites and predators, but they are useful in protecting you and helping you reproduce.

10.21228.

By containing the conflict, you delay the explosion, but you in-crease its force. The stronger the explosion, the more destructive and long-lasting the negative consequences will be. The longer the peace, the longer the war. Contradictions must be resolved instantly, and this is the essence of honesty and courage.

10.21526. Don't take the stairs two at a time.

Be simple and money will be attracted to you, because money is the simplest form of energy, it is easier for it to evolve into simple than into complex.

10.21790.

The path of love is the path of evolutionary growth, and pride is a revolution, when everything is broken and rebuilt.

10.22045. The desire for beauty.

The meaning of growth is to achieve the identity of the content of your dreams. Content evolves in order to achieve the desired form.

10.22357.

Viruses are representatives of nothingness. Inanimate, which encourages the living to evolve.

10.22421.

Human DNA it consists of 30% virus DNA. After getting over the virus and receiving immunity from it, the human DNA evolves. First, the DNA evolves, then the entire human being follows it. Thus, those who got over the coronovirus and acquired immunity to it, there is a higher and stronger stage of evolution than those who refused to evolve.

ABOUT THE AUTHOR

8.2479.

SoloINC (anc.greek "combining the uncombinable", keeper of the grain")

Soloinc Logic, philosopher from the city of Sofia. Soloinc (Diamond Solo / Solodilov Dmitry), Bulgarian psychologist and Stoic philosopher. Supporter of the merger of logical and sensory methods of cognition. He considers the connection of traditional philosophies with modern science. He is the founder of the cyber-philosophy of Syntalism (Quantum Nanophilosphy), which considers the problems of philosophy, sociology, psychology and economics in terms of systemic cybernetics and logic.

Soloinc is not the first, but the last philosopher. Evangelist and cyberpunk guru. The author of more than 73 thousand original ideas and thoughts. Main books: "Variothoughts", "Diamond Stoic", "Theory of Existence", "Money Bible", "Quantum Philosophy", "Mathematics and Progression", "Velerechie", "The Device of the Mind", "Royal Buffoon", "Liberastia" , "Surrotic", "Surfutur" and others, in total more than 888 books.

3.1753.

In fact, Variothoughts is very tedious. I have seeked the truth all my life, then I found it and concealed it in a different place. Variothoughts is an intellectual quest and a mosaic of truth, broken into thousands of pieces. I found the truth in plain sight and concealed it back as well as before... What's the point? It's a game or a way to kill boredom. We live eternally and boredom turns our life into hell. I want to save you from sufferings for some reason...

10.21128. Soloinc Music

Soloinc Music is a stunningly beautiful integrity of music and text, admiring metaphors and secret meanings. Soloinc Music is a pleasure for living minds who have dedicated their lives to the search for beauty and truth. Soloinc Music awakens the minds and ignites the heart. Everyone will find joy and strength to live in it.

10.2341. A realistic mysticism.

The genre of poetry and music of Soloinc is a mystical realism. Most Soloinc songs are mystical ballads or religious hymns, prophecies, and insights. Soloinc lyrics are always metaphors and mystical signs. They cannot be taken literally. These are grains of sand in which entire worlds are hidden. All words are the opposite. To understand the meaning of the Variothoughts texts you need to read from bottom to top, from right to left.

SYNTALISM – GENERATIVE QUANTUM NANOPHILOSPHY

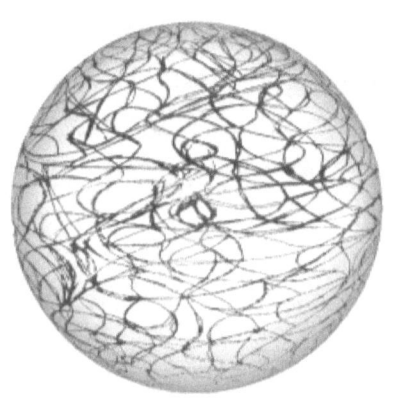 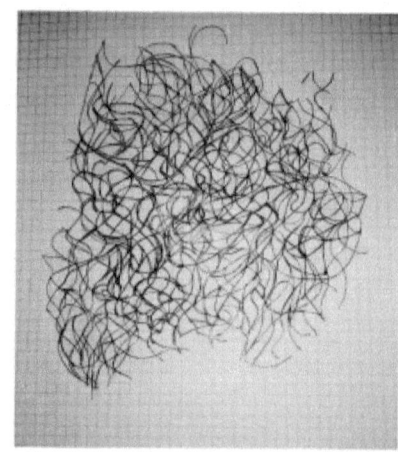

10.19296.

The philosophy of Syntalism was inspired by the poetry of life, expressed in the poems of such poets as Shakespeare, Robert Burns, Williams Blake, Pasternak, Lermontov, Mayakovsky, Velimir Khlebnikov, Paul Eluard, Andrey Bely, Alexander Blok, Voznesensky, Asadov, Gutseriev, Anna Akhmatova, Tsvetaeva, and others. Where if the philosopher had not come, the poet would have been there. Poets are like rays of light showing the way to thinkers.

5.782. Syntalism is the philosophy of the 5G generation.

Small thoughts are the philosophical system built in the millimeter wave range. Syntalism is 5G philosophy in the millimeter wave range built according to generative genetic algorithms.

5.767.

In Variothoughts, conceptualization follows the generative genetic algorithm.

5.768.

Variothoughts is the self-teaching guide on generative philosophy.

10.22348.

Syntalism is a philosophy that connects the unconnected with the goal of achieving integrity. Integrity is truth. To know the truth, the mind must cultivate tolerance and humility.

5.783.

Variothoughts is structured as a phased antenna array that ensures a dynamic horizontal and vertical growth of thought according to the generative algorithm and makes it possible to create different-sized logic data arrays. This solution minimizes energy consumed to maintain the integral information field. Variothoughts is a system of small cells in the millimeter wave (super-small thought) range in which the size of cells and their interaction structure are dynamic in nature.

10.3109. Unified system of knowledge.

The philosophy of Syntalism is by far the most perfect and clear philosophy, revealing the nature of being. Syntalism is like an ocean containing all other philosophies and religions. Syntalism understands and explains any point of view, agrees with every-one and loves everyone, considers everyone beautiful. Thou-sands of points of view, uniting into streams and rivers, turn into an ocean of Synthism.

VARIOTHOUGHTS COLLECTIBLE BOOKS

4.3423. Sand vs truth?

A book's collectable from the Variothoughts series costs only 1 cubic meter of real estate property. It is a very delicious price for something priceless.

3153.

God loves collectors as they give work to many creators...

6.6033.

The electronic version of Variothoughts is huge but printed versions are more complete and this book's collectables and handmade versions are unique in their completeness. Each of the author's gift manuscripts of Variothoughts is handmade and customized, that's why it includes even the latest texts that exist only in rough copies and have not yet been published anywhere.

10.22513.

Friends, I have not sold any Variothoughts collectibles yet. Pride rules people, that is cowardice and greed. There are very few courageous and intelligent people. He who is brave and buys the first book is very lucky. The first collector's copy of Variothoughts is a great value. Each collection book is registered and numbered. However, there will never be many of them, if I sell such books at least a few pieces a year, it will be good.